EUROPE'S FINAL FRONTIER

THE SEARCH FOR SECURITY THROUGH SPACE

EUROPE'S FINAL FRONTIER

THE SEARCH FOR SECURITY THROUGH SPACE

Alasdair McLean
and
Fraser Lovie

Nova Science Publishers, Inc.
Commack, NY

Assistant Vice President/Art Director: Maria Ester Hawrys
Graphics: Frank Grucci
Editorial Production: Susan Boriotti
Office Manager: Annette Hellinger
Acquisitions Editor: Tatiana Shohov
Book Production: Ladmila Kwartirof, Christine Mathosian,
Joanne Metal and Tammy Sauter
Circulation: Iyatunde Abdullah, Cathy DeGregory and Annette Hellinger

Library of Congress Cataloging-in-Publication Data

McLean, Alasdair W.M. and Lovie, Fraser
Europe's final frontier: the search for security through space /
McLean, Alasdair W.M. and Lovie, Fraser.
p. cm.
Includes bibliographical references and index.
ISBN 1-56072-462-5
1. Astronautics, Military--Europe, 2. Western European Union. 3. National security--Europe. I. Title.
UG 1525.E85M3497 1997 97-34374
358'.8'094--dc21 CIP

6080 Jericho Turnpike, Suite 207
Commack, New York 11725
Tele. 516-499-3103 Fax 516-499-3146
E Mail: Novascience@earthlink.net
Web Site: http://www.nexusworld.com/nova

Printed in the United States of America

CONTENTS

ACKNOWLEDGEMENTS

Inevitably, in the course of writing any book, the authors rely on the assistance of many people. We would therefore be remiss were we not to record our thanks to all those who in any way helped us in our work.

As always, however, there are a number of individuals who particularly stand out for the help or encouragement which they have given. Consequently, we wish to extend our particular thanks, for a variety of reasons, to LTC Edward Aymar, Barry Blaydes, Henri Burgelin, Kirsten Ferguson, David Greenwood, the late General Daniel O. Graham, Yves Robin, John Roper, Trevor Salmon, Mike Sheehan, Ailsa Swankie and Peter D. Zimmerman. In addition, to those others, un-named through choice (or omission on our part) who have helped in any way, we also extend our grateful thanks.

As always, credit goes to all those who have helped us write this, while all errors, whether of omission or commission, are the sole responsibility of the authors.

CHAPTER ONE

INTRODUCTION

Throughout history, military commanders have sought to exploit the high ground. Whether to engage in reconnaissance, to facilitate better lines of communication, or simply to provide a better location from which to launch an assault, the high ground is militarily important. Today, the ultimate 'high ground' is space.

The advent of satellite based military capabilities has truly changed the face of the modern battle-ground. As we approach the 21st century, command of, and the ability to make use of the space environment, has become a fundamental part in the defence planning of all the major military powers. The impact of space based assets and their ability to perform an astounding array of different tasks, whether it be in operational planning or in the areas of command and control, is now widely accepted. Indeed, it can be argued that it is now an essential component for effective military action. The Gulf War is a prime example of the advantage space-based military assets can offer, indeed that particular conflict has been hailed by many as the first 'space war'. The most significant impact of the Gulf War, in terms of its impact on future defence planning, was to bring home to low-level commanders the centrality of facilities which can only be provided via the utilisation of space in the ultimate success of the operation.

It is now 40 years since the dawn of the space age with the successful launch by the Soviet Union of the Sputnik 1 satellite. Even in 1957 the space programmes of both superpowers - the only active space participants at that time - were firmly under the auspices of the military. Today, of course, the situation is rather different, with a multitude of countries involved in civil space activity. Yet the military significance of space is now greater than ever before. Today, space is utilised by all the major military powers for a wide range of purposes, including surveillance, reconnaissance, navigation and communications, to name but four. For years, advocates of military space endeavours have been putting their case - sometimes to less than enthusiastic defence planners and politicians. Since 1991, their task has been rather easier.

Prior to this date, many had regarded space systems as an expensive luxury, unconvinced of their true utility, and in many cases not fully aware of the military advantages they could offer. Then came the Gulf War, which, for the first time, saw these systems being tested in the heat of combat. This conflict was the first in which the full panoply of space technology available to the military - in this case, the forces of the coalition against Iraq - were utilised in a combat situation.

If the Gulf War, therefore, was the first space war, what lessons can be learned from it ? While the full range of systems, number of satellites in use, and so on may never fully be known (such information remains classified) at least one estimate suggests that the total number of satellites used during the Gulf conflict was around 40[1]. Notwithstanding this, it is still possible to make an informed assessment of the value of space-based assets using publicly available information.

While clearly all the various satellite systems played their role, there is little doubt that some were of greater significance. One of the major difficulties faced by the coalition forces in the theatre of operations was that of navigating in the comparatively featureless desert terrain. This problem was overcome by making use of the Global Positioning System (GPS). Using its constellation of satellites, it can provide users with information on their precise position and velocity virtually anywhere in the world. Indeed the widespread utility of GPS was such that its use was not limited to general navigation. It was used in an airborne role to aid precision targeting, to accurately locate Iraqi minefields and routes which had been cleared through them, and to avoid accidental contact with Iraqi forces. Indeed as the official post-conflict assessment put it:

> 'Use of space-based navigation and positioning was an unqualified success. The NAVSTAR Global Positioning System (GPS) played an important role in the success of the overall operation.' [2]

But if GPS was undoubtedly everyone's favourite new item, other systems also proved their worth. The rapid build-up of forces brought about an ever-increasing demand for communication facilities, in an area of the world not over-endowed with the necessary infrastructure. Consequently, the overwhelming majority of communications requirements had to be fulfilled by way of satellite systems. At first, demand for these services far outstripped the military's capability in this field. This problem was overcome, however, by optimising the satellite constellations for use in that geographical region - in at least one case by maneuvering a Defense Satellite Communications System (DSCS) satellite from its normal location over the Pacific Ocean to a new orbit over Saudi Arabia.[3] General Colin Powell has stressed how important this feature of the war effort was:

> 'When we started our deployment, we had only the most rudimentary communications infrastructure in Southwest Asia and the challenge of distance was daunting. Thanks to good planning and to our understanding of the importance of satellites, we quickly and smoothly transitioned to a mature tactical theater network'[4]

If General Powell was impressed by the realisation of the importance of such assets, Lt General James Cassity was plainly stunned by the sheer scale of the operation taking place miles above the battleground:

> 'The services put more electronic communications connectivity into the Gulf in 90 days than we put in Europe in 40 years.'[5]

The other major areas of support which came into their own in the Gulf conflict were those of space-based surveillance and reconnaissance systems. In any conflict, timely, accurate intelligence information is a considerable asset to military commanders. In this case the American satellite surveillance and reconnaissance capability proved to be a decisive factor in both the planning and execution of operations. In addition to providing intelligence data which was used to locate and identify potential targets, satellite systems also provided one important element of the information required for damage assessment.

Although, as already indicated it is impossible to be specific about the particular satellites used due to the inevitable secrecy which surrounds their actual capabilities and effectiveness and the accompanying reluctance of Defense Department officials to discuss such issues, there is broad agreement amongst analysts as to the systems which were available. In such a conflict it is reasonable to assume that all available assets would have been used to achieve maximum benefit for the coalition forces. To give just one example

This is a clear argument in favour of the development and ultimate deployment of Anti-Satellite (ASAT) systems. While this in itself is a fascinating point, the discussion of this topic is somewhat outwith the scope of this book. For our purposes, the significance of General Kutyna's observation is the implied necessity of space-based capabilities for the effective conduct of modern military operations. The Gulf War, as the first space war, proved the importance of such systems, and pointed to the reality of modern warfare which has seen the advent of an era where the planning and incorporation of such systems into operational planning should now be regarded as essential. Furthermore, given the knowledge gained from this experience, 'the space support for Desert Storm [and] Desert Shield will probably be the minimum support expected in any future crisis,' according to Vice Admiral W. A. Dougherty, deputy commander of the US Space Command.[7] Thus it seems more than likely that the utility of space to the military forces of a multitude of states, having already grown considerably, is likely to continue to grow for the foreseeable future.

That being the case, it becomes necessary to ask how such a growth in requirement might be satisfied and how this might be achieved. Clearly in the case of the US, which is largely autonomous in this sphere, it will almost certainly continue to meet the great majority of its military space requirements on a national basis. But what of other countries - and in particular the countries of Western Europe?

Historically, the two major military space powers have been the United States and the Soviet Union/Russia. For a number of reasons, which will be elucidated elsewhere in this book, the states of Western European chose not to pursue this path with anything like the vigour of the two superpowers. Perhaps one of the most obvious reasons for this is that since 1945 the security interests of Western Europe have almost exclusively been defined within the over-arching framework of the North Atlantic Treaty Organisation (NATO). As a result of this ability to bind national fortunes into the framework of an international alliance, the security interests of the United States and Western Europe have largely been held to be indivisible. Thus, there appeared to be little point in the countries of Western Europe duplicating the efforts of their larger NATO ally. Such a course of action would simply have appeared a wasteful use of scarce resources.

However, in the ever evolving field of international security, nothing stays the same for ever. For many years, one of the most potent debates within the NATO alliance has been over the question of burden-sharing. Simply stated, there has been a constant tension underpinning the West European-American relationship which seeks to assess the extent to which the Western European partners should be responsible for the burden of their own security. Today, this debate has moved on somewhat from the relatively static position it occupied throughout much of the 1970s and 1980s. In the 1990s a more mature debate has evolved, looking to move the organisations responsible for the security of Europe into the next century. As a consequence, the notion of burden-sharing has largely been overtaken by the widespread acceptance of the need to establish a semi-autonomous European Security and Defence Identity (ESDI). In 1991, for example, the Western European Union (WEU) Council of Ministers noted that an ESDI was emerging within the Twelve [members of the European Communities] and WEU. The emergence of such an ESDI was seen as being a positive enhancement of the existing alliance framework, with the Council agreeing:

> '... that the development of a genuine European security and defence identity will be reflected in the strengthening of the European pillar within the Atlantic Alliance'. [8]

All at once it had become apparent that there could be a different relationship between the states of Western Europe and the United States of America, a relationship that remained within the framework of the Atlantic Alliance, but which accepted somewhat changed roles for the partners, a relationship which began to appreciate that in the ever-changing world of international relations, the security interests of the US and Europe may not always be synonymous.

If this process of redefinition was to be successful - if it was to strengthen, rather than weaken the long standing and successful security relationship which bridged the North Atlantic - then Europe had to aspire to autonomy in areas which hitherto had been the domain of its North American allies. Perhaps the single largest area in which the European states were lacking in this respect was to be found in relation to the military use of space. Thus, it can be argued that, if a true ESDI was to be created, Europe would have to examine its requirements in this area, and move in the direction of at least a partial meeting of those requirements on its own.

The 1990s have thus developed into the decade which, from an operational military perspective, has seen space become Europe's final frontier.

Bearing all of these points in mind, it is the aim of this book to examine Western Europe's response to this challenge by tracing, in parallel, the evolution of political and strategic thinking which has resulted in an increased desire for greater European autonomy in the field of security and defence, and the detail of the manner in which it has tried to move towards the attainment of this increased autonomy in the specific area of military space capability. In so doing, considerable attention will be focused on the WEU as an organisation - the only one of its kind in Europe which has, over many years, devoted a significant amount of attention to the field of military space. This involvement in the evaluation of space as a means to further security has allowed the WEU to develop a high level of competence in this area, something which has seen the organisation play a key role in the development of European military space activity. Attention will, of course, also be paid to other multi-national, or national efforts which combine to form a part of Europe's search for security through space.

Chapter Two examines the development of the Western European Union. It traces the origins of this often-overlooked security organisation - which in fact predates NATO - and examines its early promise and subsequent decline as an effective organisation. In so doing, it locates the WEU within the traditional security architecture of Western Europe - the transatlantic security relationship which has persisted since the end of the Second World War.

Throughout the lean years for WEU, one of the significant areas which it, almost alone, continued to maintain an interest in, was the area of European space policy, with particular emphasis placed on the development of European military space policy - or more frequently the lack of it. Chapter

Three examines in some depth the work of this organisation in relation to space, tracing the numerous reports and recommendations which it produced on the topic from the early 1960s up to its revitalisation as an organisation in the early 1980s. Throughout these 20 years, space was a topic never far from the attention of the WEU Assembly, and its work in this area can be legitimately regarded as laying the groundwork for the present situation, in which active consideration is given to the development of a European military space policy.

The most concrete example of this is the establishment by WEU of a Satellite Centre in Torrejon, Spain. Consequently, Chapter Four examines the moves which resulted in the establishment of this Centre and its early days of operation.

It would, however, be entirely incorrect to assume that the only military space programmes in Europe fall under the auspices of the WEU. A number of countries, such as France and the UK, have, for many years, operated military satellite systems. Chapter Five, therefore, offers a detailed assessment of the military space capabilities currently operated by the European states, and also goes on to outline some of the future programmes to which they are already committed.

Developments in military space capability cannot, however, be regarded in isolation. They are but one element of the overall arrangements which must be in place to guarantee European security. Since the end of the Cold War, there has been considerable debate as to the relevance and development of these structures. Chapter Six, therefore, examines the post-Cold War security architecture of Europe, paying particular attention to the rapidly growing and changing role of the Western European Union within this framework.

In Chapter Seven, the tack changes, from an investigation of the historic framework of European military space and instead attempts to divine the future of that most crucial of military support activities. In this chapter attention is devoted to the possible way in which European military space cooperation may develop over the next 20 years or so. It examines a number of different options and routes for the effective, efficient development of a specifically European capability. However this may develop, it is to be remembered that it will only do so within the broader context of the development of the whole field of European security. This is the focus of Chapter Eight, which concludes our study and moves from the specific field of military space policy and its possible development as identified in the preceding chapter, to examine the whole question of the way in which the overall security architecture of Europe may develop in the years ahead.

Thus it is hoped that this book serves two purposes: while focusing on the specific field of military space activity as a distinct issue area, it also traces the development, growth and re-emergence of the WEU as a European security organisation, all the while placing both of these within the broad framework of European security in general.

It is essential that Europe considers this area in significant detail. Any pretension at independence of action in the military sphere will be predicated upon the ability of Europe to provide the full range of command, control and communications for its forces. At this time, as the security framework which will cover the continent in the next millennium is being shaped, it is vital that Europe embrace the reality of a situation which dictates that a co-ordinated military space policy will be an essential component of future successes.

ENDNOTES

[1] *Space News*, Vol. 2, No 16, May 6-12, 1991, p.6

[2] US Department of Defense, *Conduct of the Persian Gulf War*, Final Report to Congress, (US Department of Defense, Washington D.C., April 1992), p. K-41

[3] *Aviation Week and Space Technology*, May 13, 1991, p.89

[4] Powell, General C., December 1990, quoted in *Conduct of the Persian Gulf War*, Final Report to Congress, (US Department of Defense, Washington D.C., April 1992), p.K-30

[5] Cassity, Lt. General J.S., quoted in *Conduct of the Persian Gulf War*, Final Report to Congress, (US Department of Defense Washington D.C., April 1992), p. K-26

[6] *Space News*, Vol. 2, No 16, May 6-12, 1991, p.6

[7] *Space News*, Vol. 2, No 13, April 15-21, 1991, p.4

[8] WEU Council of Ministers, *Communique*, (Vianden, Luxembourg, 27 June 1991) p.2

Chapter Two

The WEU: History and Development

I. Introduction

This chapter places the WEU in to the historical context of the post Second World War security structures. This analysis will look at the relatively enfeebled beginnings of this organisation and the process of development, stagnation and rebirth which it has experienced. In order to do this the creation and early years of the WEU will be traced, examining the rationale behind the organisation and its ostensibly political involvement in the security debate of the 1950s, before charting its extended period of inactivity in the 1960s and 1970s which preceded the organisation's eventual rebirth in the 1980s.

This chapter will therefore help to place in context the remarkable change in fortunes the WEU has enjoyed since the mid 1980s, when the decision to re-evaluate and utilise the structure which existed for West European defence cooperation was taken. Thus it allows us to examine the expansion of membership and the inclusion of partner countries, as well as the development of an operational capacity, both of which are cited as clear evidence of the commitment to a European defence and security system, although as will be seen, there is an obvious desire to ensure that this is not to the detriment of the Atlantic linkage and the notion of interlocking institutions.

II. The Creation of the WEU

In the immediate aftermath of World War II, the security situation in Europe remained in a state of flux. The United States proved unwilling to provide the kind of security blanket that many European states desired, unmoved at this stage by the necessity for such a commitment following their decisive involvement in the Second World War. However, the Europeans saw American involvement in their security as a crucial element in their efforts to adapt to the political and economic circumstances created by the disruption of the war and the damage it had done to the infrastructure of the continent. While the immediate concern of the states was their economic well being, the security imperative impinged on their ability to concentrate exclusively on economic regeneration. In the corridors of power in the United Kingdom and, especially, France, an ever present consideration in the emotionally charged atmosphere of post 1945 Europe was their residual fear of Germany, coupled at that time to an inability to fully comprehend the role that the Soviet Union sought to play in post War European politics. The reluctance of the United States to remain a stabilising force in Europe helped forge a consensus that some form of European security institution, albeit limited in nature, was necessary and that the conditions for some attempt on behalf of the European states themselves to provide for their own security were evident.

It was as a consequence of this realisation that the incremental development of the post-war European security system had as its first act, the signing of the Anglo-French, Dunkirk Treaty in 1947. This bilateral Treaty allowed for economic and security cooperation, with the fear of a renaissance in German power serving as the major focusing principle, although it might also be regarded as the first, tentative step towards common action against the emerging threat of the Soviet Union.

The notion of close cooperation between the British and French was in itself nothing new. In 1940, the British Prime Minister, Winston Churchill had aired the possibility of a conjoining of the two states in a framework where there would 'no longer be two nations, but one Franco-British union.'[9]

Although this suggestion was to be ignored by the French, and despite the fact that it can be viewed with a certain level of retrospective mirth given the attitude much of today's Conservative Party has towards 'Europe', the notion of integrative action in the face of adverse circumstances had been publicly aired. Notwithstanding the fact that what was eventually to follow by no means matched the depth of action that this early idea had envisaged, the basic premise of the resultant defence agreement was the same: the preservation of the security of both nations through cooperation.

While the Dunkirk Treaty was little more than a signalling of resolve on behalf of the two war time allies, this agreement between the two major European powers was to become the framework for a broader effort the following year. With the signing of the extravagantly titled, 'Treaty of Economic, Social and Cultural Collaboration and Collective Self-Defence', more commonly referred to as the Brussels Treaty, the principles at the heart of the Dunkirk document were embedded with the inclusion of additional members and a broadening of the Treaty's remit. The addition of Belgium, the Netherlands and Luxembourg (the Benelux states) and the expansion in scope which the new document provided, with the inclusion of elements relating to social and cultural exchange, made the Brussels Treaty a much more wide-ranging document than its Anglo-French predecessor.

Signed in the Belgian capital in March of 1948, this five nation agreement was designed to allow for collective action in the face of aggression and was also to serve as the embryo from which the Western European Union would eventually flourish.

The central tenet of the Brussels Treaty was without doubt the notion of collective self defence contained in Article 4, which most clearly outlined the obligations of the signatories in the event of external aggression:

> 'If any of the High Contracting Parties should be the object of an armed attack in Europe, the other High Contracting Parties will, in accordance with the provisions of Article 51 of the Charter of the United Nations, afford the Party so attacked all the military and other aid and assistance in their power.' [10]

This clearly emphasises a high level of compulsion on the part of the membership to support actively any of their colleagues in such a time of crisis. This is something which was, as we shall see, lacking from the Washington Treaty which would subsequently outline the obligations of members within the framework of the North Atlantic alliance.

With this military and security element firmly embedded, the remainder of the Treaty sought to expand upon the dominant elements that had characterised the Dunkirk Treaty, namely defence and economic assistance, through the inclusion of clauses designed to enhance social and cultural linkage across the partner countries. The Brussels Treaty thus sought, through its Articles 2 and 3, to promote the development of mechanisms for the preservation of social security and the attainment of a higher standard of living across Western Europe as well as highlighting the need to better understand the cultural 'principles which form the basis of their common civilisation', something which it was understood would be achieved partly through the promotion of cultural exchanges.[11]

The inclusion of these elements and the clear intention to move the organisation beyond considerations of a purely defence and military nature can be taken as evidence of the part that the Brussels Treaty had in a broader vision of European cooperation and as part of an alternative view of how best security and peace in Europe could be achieved. Despite the clear military overtones of the Treaty, the evidence points to an acceptance of a definition of security which extends beyond the military and into areas such as economics and culture. That the WEU was eventually to play a minimal role in the provision of any of the functions which the Treaty envisaged, had more to do with the existence of other international organisation better equipped to expound and enact them, than it had to do with any particular flaw in the vision of those who drafted the Treaty. Indeed, one of the most directly significant impacts of the Brussels Treaty was that it helped to establish conditions within which ongoing international collaboration could be generated in a range of different areas.

Yet, despite according the Brussels Treaty this visionary status, it is clear that the Treaty's ultimate objective and the rationale underpinning both it and the Dunkirk Treaty before it, was the pursuit of defensive agreements designed to bind allies into collective defence arrangements. More than that, they were designed with a very specific threat in mind. The fear of a resurgent German challenge to the security of Europe was tangible and the Brussels Treaty made no attempt to camouflage this, referring directly to the fear the member states had that Europe might be forced:

> 'To take such steps as may be held to be necessary in the event of renewal by Germany of a policy of aggression.' [12]

Such a position was not limited to the states of Europe, with Germany's role as an aggressor in World War Two alluded to very strongly in the Charter of the United Nations which states:

> 'The term enemy state ... applies to any state which during the Second World War has been an enemy of any signatory of the present Charter.' [13]

Given the prevailing circumstances in the post-War period and the all too well remembered lessons learned in the aftermath of World War I, where the creation of the Weimar government had helped mask the residual potential for militancy in German domestic and foreign policy, this reaction can perhaps be deemed an understandable approach to uncertain times. However, despite the desire to punish Germany, an attitude most clearly prevalent in France, it soon became apparent that the best course of action for Europe was not to leave the Germans (and to a lesser extent the Italians) in the international wilderness, but was instead to utilise some mechanism through which both states could be effectively re-integrated into the fold of responsible European nations and thus allow them to resume a role within the broader scope of international society.

The status of Germany therefore proved a constant concern, with the post-War settlement which had seen the country split into four military zones of control, each under the jurisdiction of one of the victorious World War II states,[14] considered by many to serve none of the parties best interests. There was an obvious need to define the role that this crucial part of the European landmass was to play in the geo-politics of the continent.

The necessity to push this agenda forward became steadily more evident as the situation in Europe developed following 1945. While the legacy of the War had left its mark upon the framers of the Brussels Treaty, the most pressing security concern of West European states by the late 1940s was no longer whether the Germans would once again seek to assert their influence on the continent but rather the principle focus of attention had shifted to the developing Cold War with the Soviet Union and the fear of an ideological gulf that the capitalist 'West' could not, and would not, bridge. As Winston Churchill famously informed us in 1946, an 'Iron Curtain' had indeed descended across the continent of Europe and, notably, much of Germany was firmly on the Western side of it.[15]

The harsh reality of the security situation at this time was that a far more tangible threat to the security of Western Europe was perceived to exist, that of Communism and the ideological impasse that was taking shape, as the fundamentally opposing ideas of the capitalist, liberal-democratic West, rubbed shoulders with the authoritarian, command economies of the Soviet bloc. Ideological paranoia and the security spectre which it raised was beginning to replace the fear of latent Nazism as the defining feature of Western security planning, with the most public evidence of this 'new' challenge coming in the form of the Soviet blockade of West Berlin in 1948.

Following what was seen in the West as Soviet sponsored adventurism in both the domestic political situations of Greece and Czechoslovakia, with Soviet involvement in both the Greek civil war and interference widely condemned as anti-democratic following their involvement in the politics of Czechoslovakia at a time when the Prague government was taking shape following the war, the Soviet attempt to force the Western powers to abandon West Berlin proved to be a decisive moment in the direction of West European security, indicating Western Europe's vulnerability and thus its need for American support to ensure security.

The blockading of road, rail and canal linkages into West Berlin, left the allies with only one option in their attempt to supply the beleaguered city: the provision of a huge airlift operation, which was to last from June 1948 to May

1949.[16] Not only did it help prove the Americans' commitment to the continent, it spurred the formalisation of the de facto political consensus that already existed among most of non-communist Europe and thereby helped prompt the decision to create the North Atlantic Treaty Organisation and to push forward with the rehabilitation of occupied Germany, a process which was further advanced through the creation of the Federal Republic of (West) Germany. Thus it can be seen that the spring of 1949 was to prove a pivotal time in the creation of the post World War Two structure of Europe.

As it became gradually more obvious that the situation vis-à-vis the Soviet Union was unlikely to develop into one of amicable relations, the British, French and Americans abandoned attempts to 'solve' the German question and decided to cede partial sovereignty back to a West German civilian government. From September 1948 through to May 1949, a newly formed Parliamentary Council in Bonn deliberated over the form of the Basic Law, which would serve as the constitutional basis for West German sovereignty. Following several months of negotiations the Federal Republic of Germany (FRG) came into existence on the 23rd of May, 1949. Following parliamentary elections held later in the year, the voters elected the FRG's first government and Chancellor, with the Christian Democrats emerging as the largest party and Adenauer becoming the first post-War Chancellor. This gave the FRG a legitimacy which the divided Germany had otherwise lacked and prompted an almost immediate reaction in the Soviet controlled zone, an area which was later to become the German Democratic Republic (GDR) in October of 1949. This bifurcation of Germany effectively ended any remaining hopes for a rapprochement between East and West, and instead the division of Germany was eventually to provide the most symbolic feature of the Cold War, when in August, 1961 the East German authorities erected the Berlin Wall in a bid to stem the flow of refugees from the East to the West.

The signing of the North Atlantic Treaty in April 1949, brought a transatlantic element into the security of Western Europe and helped to allay some of the concerns surrounding the inability of the West European states to effectively defend themselves alone. This 'Atlantic Alliance' has served as the basis for West European security since that day and is currently engaged in an extended period of transition as a consequence of the altered strategic circumstances of the 1990s.

Yet despite the durability of the alliance, the security guarantees contained within the North Atlantic Treaty are by no means an unequivocal agreement to ensure collective action in the face of aggression, although in practice it was always understood that the provisions would oblige the United States and the other members to become involved if any of the member states became the subject of unwanted attentions from an external source. Thus, Article V of the Washington Treaty states that:

> 'The Parties agree that an armed attack against one or more of them in Europe or North America shall be considered an attack against them all and consequently they agree that, if such an armed attack occurs, each of them, in exercise of the right of individual or collective self-defence recognised by Article 51 of the Charter of the United Nations, will assist the Party or Parties so attacked by taking forthwith, individually and in concert with the other Parties, such action as it deems necessary, including the use of armed force, to restore and maintain the security of the North Atlantic area.' [17]

This contrasts with the Brussels Treaty, upon which the collective defence commitment of the North Atlantic Treaty was loosely based, and which unquestionably guarantees the involvement of the members should one be the subject of an aggressive action. The obvious caveat here is that without the backing of significant military forces such a guarantee is relatively worthless and while the combined efforts of the French, British and their European allies may well have been considerable, the need to have a broader based security structure in place to counter-act the sheer size and power potential of two major actors (in the shape of Germany and the Soviet Union) was obvious. With the British and French forces already stretched as they struggled to maintain their pretension to world power as it buckled under the pressure of the decolonisation process, while at the same time attempting to press forward with the process of developing their fragile economies, there remained a fundamental inability on behalf of Western Europe to defend itself credibly.

Once again it can be seen that the role of the United states was of critical importance. Its willingness to aid the economic reconstruction of Europe through the Marshall Aid programme, was to be mirrored in the security realm by its foreign policy stance espoused in the Truman Doctrine. In 1947, President Truman had famously enunciated the promise of assistance for those peoples who wished to protect themselves from tyranny. Largely a direct response to the events which had taken place during the Greek civil war, this also served as the basis for anti-Communist intervention on a global basis. The necessity for such a doctrine became immediately relevant in a European context, when, in 1948 the Finns resisted pressure to enter into a treaty with the Soviet Union similar to those which had been imposed on the countries of Eastern Europe.[18] That the Norwegians and Danes felt threatened enough by such developments to mobilise their reserve forces, suggests that the perceived threat was a real one.

The onset of the Korean War served as additional evidence of the value of the Truman Doctrine and provided a clear example for those who sought to demonstrate the potentially expansionist tendencies of communist ideology. Coupled to the less than perfect guarantees included in the North Atlantic Treaty framework, these developments placed the issue of the new Federal Republic's status within European security very much back at the top of the agenda.

The American involvement in Korea and an inability or unwillingness on behalf of the West European states to spend valuable resources on the military in a period of strained economic recovery, led to the logical conclusion that West Germany would have to play a part in the defence of Europe. At this stage the sovereignty of the FRG was still a moot point. It was a state with no defence capabilities and an all too apparent willingness to fulfil the role of dutiful neighbour, something which was borne of a deeply felt guilt regarding German conduct during the War. Although this gave the West Germans an opportunity to establish themselves as a responsible part of the international system, something they were to achieve with aplomb, it did nothing to help solve the underlying security dilemma that a divided and non-military

Germany prompted. While it was hard for states which had been on the wrong end of the Nazi war machine to accept the gradual integration of the new West Germany back into the life of the continent, an emasculated giant at the heart of Europe did nothing for the overall security of the area.

As this issue became gradually more acute, the United States took the diplomatic initiative and tabled a suggestion that the rearmament of West Germany begin. This was understandably received with muted support in the capitals of Europe, whose foremost security consideration remained the need to guarantee the acquiescence of West Germany. A relatively brief five years since the conclusion of the war, made September 1950 appear just a little too soon for many to contemplate such an option.

As a means of avoiding this apparent excess, an alternative solution emerged with the first steps taken towards the creation of a European Defence Community (EDC). Based on the ideas of the French Prime Minister, René Pleven, and following the same basic blueprint as the European Coal and Steel Community (ECSC) - which had come into effect in April 1951[19] - the Pleven Plan envisaged the creation of a European army under a European Defence Minister. It was proposed that this organ could be responsible to a West European parliament, and as such this plan sought to promote the dual goals of broader integration in Europe and just as importantly, to provide a mechanism for the involvement of West Germany in the defence of Western Europe.

Over the next four years the EDC followed a tortuous route, pitting those who sought the rearmament of West Germany in a NATO context against those for whom the notion of an EDC was perceived as being the preferable option. Despite the crucial nature of this discussion, the British Government refused to become involved in what it deemed a continental European venture, a factor which, in the minds of many senior French officials, undermined the entire project before it had even begun. When the EDC Treaty was signed in Paris in May 1952, the French, Italians, West Germans and the Benelux states theoretically agreed to push forward the notion of a European army that the original EDC plan had outlined. However, open hostility from the French parliament meant that the Treaty was never put before the National Assembly for discussion, let alone ever coming close to full ratification. With this rejection on the part of the French to entertain such a concept the entire EDC project was effectively stillborn. When the French subsequently suffered yet another embarrassing military reverse in 1954, this time in Indo-China at Dien Bien Phu, it merely confirmed the fear that any plan which did not attract wholehearted political involvement from the United Kingdom and which was not further augmented by a meaningful set of military commitments, was simply not credible. Without British forces, or at the very least the guarantee of their use if required alongside the French, it was felt that the resultant structure would have placed too much emphasis upon the West German element and was therefore not something that the basic political and military sensibilities of the time could ever allow. The French were quite understandably unwilling to allow their security to become dependent upon cooperation with their traditional enemies.

This left Western Europe in virtually the same situation as it had been prior to Pleven's introduction of the EDC notion: West Germany still had to be integrated into the defence of the continent; the United States was still advocating a policy of re-armament; and the British were still reluctant to commit to anything which could be classified as a predominantly continental European concern. The one forum which appeared to offer an avenue through this impasse was the Brussels Treaty. As the only security oriented agreement with a solely West European membership and which included both the French and British among its membership, the Brussels Treaty ensured that one of the crucial features of any solution was in place: namely that the British and the French be actively involved. Although it had to a large extent been usurped by the broader membership of NATO, here lay a means of getting both the major West European powers involved while at the same time allowing the United States to cede to Western Europe the authority for finding a solution to the German problem.

The Brussels Treaty therefore offered a compromise: allowing the newly sovereign states of West Germany and Italy to take their places in the security arrangements of Western Europe, while at this stage not necessarily holding out the guarantee of full integration into the broader structures offered by NATO.

As a result, the Paris Agreements of 1954 amended the Brussels Treaty to allow for West German and Italian membership of the re-titled Western European Union. This was to be the forum through which West Germany could be integrated into the military aspects of the international community and play a role in the security concerns of the continent. Whereas the original Brussels Treaty and for that matter the London Treaty had specifically cited the Germans as the focus of the security guarantees contained therein, the amended Brussels Treaty sought to redress this with the replacement of references to the fear of resurgent German power with the far more suitable notion that the Treaty was designed:

> 'To promote the unity and to encourage the progressive integration of Europe.'[20]

With the Federal Republic now accepted as a valuable part of the Western alliance and with certain features in place to help pacify fears of an unchecked West German involvement - notably a United Kingdom commitment to armed forces on mainland Europe, the lack of which had damaged the progress of the EDC, and an ongoing desire to limit the heavy armaments capacity of West Germany,[21] - the accession of West Germany and Italy to NATO in 1955 was merely the final act in the formalisation of this new position and evidence of the return to full sovereignty for those countries.

The necessity of having a series of Western security organisations such as NATO and the WEU, seemed only to be confirmed by a catalogue of events in 1956. In response to the creation of NATO, and more specifically as a result of West German integration into the organisation - something which had been strenuously opposed by the Soviet Union - the Soviets established the Warsaw Treaty Organisation, more commonly referred to as the Warsaw Pact. This effectively recognised the state of East Germany and bound it and the other East European satellite states into an organisation which had Moscow firmly at its epicentre. When the Soviet Union backed the Hungarian anti-liberalisation forces of János Kádár during November of that year, it became ever more apparent that the bifurcation of the continent was complete.

By the end of 1956 the principal actors in European security appeared to have clearly delineated roles for themselves and set the scene for a production which, with only minor changes in cast and scenery, would run for over thirty years. For the WEU which had played a crucial part in the prologue, helping to characterise involvement of the FRG and Italy in this epic, it was to be a frustrating time, much of which was spent waiting in the wings of European security for its opportunity in the limelight.

III. Organisational Decline

Despite the impressive pedigree of the WEU and the centrality of the role that it played in the solution to one of Europe's most intractable political and military problems, the first thirty years of its existence merit little more than a footnote in the annals of post-War European history. The organisation was essentially a governmental construct, designed to facilitate the integration of West Germany back into the state system and the gradual diminution of the role that the WEU played after this initial success was testament to the fact that the membership had little further use for it.

This disregard for the WEU is amply demonstrated by a cursory glance at the organisation's track record in the three decades following its inception. The broad ranging vision that the original Brussels Treaty encompassed, had resulted in the signing of a multi-functional document which served as the basis for economic, social, cultural and most importantly military cooperation between the member states. If we are to judge the progress of the WEU, it is reasonable to examine the organisation on the input it had to development in these areas and most importantly, it is worth while looking at the central feature of the organisation, the commitment to collective self-defence.

On the positive side of the ledger, the WEU can rightly claim that it was from within the Brussels Treaty vision of a collective-defence structure that the blueprint for much of what followed in the realms of European security guarantees can be found. Where Article 4 of the 1948 Brussels Treaty had set forth the notion of collective action among the members in the case of attack, we have already seen that the Washington Treaty included a paler imitation of this guarantee as the basis for the security guarantee shared amongst the NATO membership.

However, the robust nature of the security guarantee contained in the Brussels Treaty ignores one of the key concepts in any such alliance, that of credibility. While the NATO membership commit themselves to much less in terms of immediate response to crisis situations, it is a more impressive lever than the Brussels Treaty could ever muster because of the involvement of the United States. The lack of credible military might to bolster such a guarantee meant that the WEU effectively ceded its role as a collective defence organisation to its more muscular cousin, NATO. The eventual decision to absorb the military structure of the Western European Union into NATO and to avoid duplication of effort among the two organisations, with many of the political and military delegations at NATO being dual-tasked and serving a similar function for the WEU, it can be seen that the WEU effectively handed primary responsibility for defence of the continent to the larger organisation.

While this military abrogation seriously damaged the credibility of the WEU, a parallel set of developments saw the economic, social and cultural roles originally envisaged by the Brussels Treaty gradually superseded by the expansion of cooperative linkages through other international organisations. The ongoing process of decline can thus be further traced through a series of transfers of responsibility. These transfers ranged from the decision to hand over the remnants of the cultural and social roles, which had been so prominent in the original Treaty to the Council of Europe,[22] and include the patent duplication of effort that would have occurred had the WEU sought to maintain an economic role, something for which the European Economic Community was clearly better placed and more capable of handling.

For all of these reasons the initial promise of the organisation dissipated relatively quickly and the WEU became largely extraneous to the requirements of the member state's governments. It was neither active nor relevant and offered nothing that was not being provided by some other of the panoply of international organisational affiliations that the member states enjoyed. When one considers the resource implications which multiple organisational memberships entail, and place this in the context of a period where an emphasis was placed upon the process of economic regeneration, it is unsurprising that the WEU found itself sidelined.

Notwithstanding these serious setbacks the organisation remained alive, if somewhat subdued, throughout this period and could claim the distinction of being the only solely West European body tasked with the remit of discussing security and defence matters. Given that NATO's North American connections and the economic remit of the EC meant that neither had the ability to provide this service, this task is one which the WEU was able to exploit as some kind of life support mechanism. Through its parliamentary Assembly the WEU was the only organisation capable of providing a discussion forum for such important issues at a non-Ministerial level, and as a result (as shall be developed in subsequent chapters) the Assembly has found itself frequently the reporter and critic of European activities in the defence and security field, in particular with relation to the exploitation and utility of space as a military asset.

Yet despite this apparent potential, the primary role played by the WEU throughout much of the 1960s seems to have been the provision of a conduit through which a functional linkage between the British government in London and 'the Six' signatories of the Treaty of Rome could be maintained.[23] When this minimal role was taken away from the organisation by the United Kingdom's accession to the European Economic Community in 1973, the WEU slipped into an apparently terminal state of disrepair. The early 1970s also saw the development of the concept of European Political Cooperation (EPC), with the Davignon Report of 1970 suggesting that the best way for the European Communities to move the process of further integration forward, was to concentrate on developing areas where a high level of common interest already existed, with foreign policy cited as a prime example.[24] With this move towards discussion of foreign policy issues within the broad framework of an inter-governmental subset of the EC, the remaining merits of the WEU as any kind of serious forum had clearly suffered yet another major blow. The subsequent suspension of Ministerial activity (the WEU did not meet at Ministerial level again until 1984) further suggested that the member states had opted for a policy of voluntary euthanasia, dispensing with the last real vestige of influence that the organisation had clung on to.

Yet despite this near death experience, the WEU did not disappear completely from the scene, remaining alive, at least in the pages of international law. Indeed, arguably one of the key elements which permitted the organisation to continue its existence was its Assembly. The parliamentarians from each of the member states who comprised this body, valued its role as a unique forum in which they could discuss matters such as defence and security. It afforded them the opportunity to meet with their counterparts from other states and to address matters of common concern. Despite the downgrading of the organisation as a whole, the Assembly of the WEU continued to meet throughout this entire period, to discuss and through its committees, to produce reports, on a range of issues appertaining to questions of importance in Western Europe.

IV. The Reactivation of the WEU

The prolonged period of disuse which the WEU suffered and the clear lack of any realistic desire among the membership to reverse this situation, could quite easily have resulted in the organisation's demise. Indeed, had it not been for developments in the Cold War during the first half of the 1980s, the WEU would, most probably, have been forgotten about and allowed to fold when the Brussels Treaty comes up for reconsideration in 1998.

The initial prompt for reactivation of the WEU came from a fear of the renewed Cold War, something which had ensued as the result of a breakdown in détente during the first phase of the Reagan administration. Several factors had helped prompt the investigation of an alternative to the NATO and EC framework which served as the primary international outlets of discussion for the states of Western Europe. Three key issues gave the WEU the impetus required to embark on the road to recovery. The first was the failure of the 1981 Genscher-Colombo plan, a bold attempt to develop the EPC into a forum for the discussion of security and defence matters.[25] In addition to this European failure, came evidence of the United States overwhelming propensity to place self interest ahead of alliance unity, with the unilateral announcement of the Strategic Defense Initiative (SDI) giving rise to a tangible fear of the destabilising effect which it was felt such a scheme would have on the delicate balance of nuclear deterrence in the continent. There had also been a lengthy, and at times acrimonious, debate on the 'Euromissile' question; the desire, or necessity for United States deployment of nuclear missiles on European territory, which, while regarded as essential by defence experts, was vehemently opposed by some sectors of public opinion, most notably in West Germany and the United Kingdom. As a consequence of these issues, a feeling had developed that the Europeans were being ignored by the superpowers and had no real input to the decision making process regarding the creation of alliance security and defence policy.

Thus the revival of the WEU, after thirty years of relative inactivity, had as its impetus the burgeoning realisation among West European states that a more coherent approach was required to discussion of their security interests. While the main forum within which discussion of security matters was, and remains, NATO, the decision was taken to reinvigorate the WEU. There was, after all, no reason to believe the Americans would, or should, be the ultimate guarantor of their security in perpetuity, and there was a growing realisation that European security interests may not necessarily always coincide with those of the United States.

In the period from 1984 to 1987 a series of Ministerial meetings brought the WEU back into the fold of active organisations and sought to legitimise its role through the provision of a coherent security forum for the members, all of whom were part of both the EEC and NATO.

The first stage in this process came with a meeting of Foreign Ministers in Paris in June of 1984, the first Ministerial level meeting of the WEU since the UK's accession to the European Communities. That meeting set the scene for a conference of Foreign and Defence Ministers later that year which resulted in the Rome Declaration, described by the former Secretary General, Alfred Cahen, as the 'rebirth certificate' of the organisation.[26] The parties agreed to reactivate the WEU and provide it with the capacity to operate alongside the other organisations involved in the provision of security in Europe. Thus the member states underlined their determination to make better use of the WEU framework, something they hoped to achieve by:

- strengthening peace and security

- promoting unity and encouraging progressive integration
- cooperating closely with members of other organisations [27]

Making better use of the framework was not something that was particularly hard to achieve or agree to, with a ten year period of stagnation suggesting that any attempt to implement WEU structures could be viewed as a success. The Rome Summit set in motion a process of refurbishment, which has culminated in the organisation's ability to claim a place at the heart of the debate over the expansion and future of European integration and the nature of the security commitment that the states share.

While this initial phase in the reactivation process had come as Cold War tensions resurfaced, it was ironically the first tentative steps towards a thawing of the Cold War that most clearly highlighted the need for a European approach to security matters and which served as the backdrop for further developments.

The Reagan-Gorbachev summit in Reykjavik (1985) and the announcement of the intention to remove the intermediate class of nuclear weapons (INF), left the European partners of NATO in a pronounced state of anxiety. They firmly believed that these weapons served a crucial role in both the deterrence of the Soviet Union, and in the inevitable escalation ladder that would guarantee US involvement should the unthinkable happen and Europe become the battlefield for a nuclear war. With the revelations of Reykjavik it became apparent that the United States and the Soviet Union were more interested in their own bilateral superpower dialogue, than they were in the sensibilities of those whom their decisions directly affected. It was therefore no surprise that this move was looked upon unfavourably by the European partners in NATO, whose perception was that a major shift in the military balance in their region had been imposed upon them without prior discussion or consultation.

That the West European states had remained blissfully unaware of this shift in the negotiating tack of the superpowers only served to promote the sense of injustice and desire to force forward a European security agenda. Moreover, this episode served to weaken the belief in the enduring nature of the American security guarantee, something that was already a constant source of concern for the far weaker military establishments of the West

European powers. It was this high-handed attitude towards world security and the lack of consideration which prompted some Europeans to begin to accept the logic of a WEU which could be seen as a legitimate forum for the discussion of West European security concerns - one that could operate independently of the United States, and which need not rely heavily on the predominant partner in NATO.

The next stage in the process came with the Hague Platform of 1987. The Ministerial meeting produced a 'Platform on European Security Interests', which is seen as the 'culmination of the first phase in WEU's reactivation'.[28] The primary purpose of the 'Platform' was to enhance the reputation of WEU as a vehicle for the pursuit of West European defence and security interests, while maintaining linkages with NATO. As a result the member states pointed out that:

> 'It is our conviction that a more united Europe will make a stronger contribution to the Alliance, to the benefit of Western security as a whole. This will enhance the European role in the Alliance and ensure the basis for a balanced partnership across the Atlantic. We are resolved to strengthen the European pillar of the Alliance.' [29]

As part of its gradual process of development, the WEU clearly had to move beyond the limited membership which it possessed, something which neither represented the full complement of the European group within NATO, nor barely more than half of the then membership of the EC. For the WEU to gain some realistic leverage and a level of legitimacy, this situation had to be reversed and the organisation was obliged to consider the issue of admitting new members. Again the Hague meeting provided impetus for this element in the growth of the organisation, setting the enlargement process in motion by opening accession negotiations with Spain and Portugal.[30] The first stage in this enlargement was the eventual introduction of the Iberian states in 1990, which has subsequently been followed by further changes in the traditional membership, with the Greeks joining as full members, and through the introduction of several different types of membership designed to allow for the incorporation of states for whom the option to participate as full members is deemed impossible, either because of their own political or ideological convictions, or because they are deemed economically, militarily or otherwise incapable of fulfilling their obligations as full members.

The gradual invigoration of the WEU stepped up a gear with the onset of the 1990s. The most obvious boost to the organisation's fortunes came with the signing of the Maastricht Treaty on European Union and the explicit role envisaged for the WEU in helping develop a Common Foreign and Security Policy (CFSP). Maastricht helped define a vision of European security for the next century and although the notion of a CFSP remains at this early stage an inter-governmental process, the scene has been set for a possible unified approach to security and foreign policy. This move has also succeeded in broadening the potential scope of the European Union's activities, for the first time openly acknowledging the legitimacy of an involvement in issues of defence and security. With the security of the continent increasingly seen as indivisible from the other arenas in which the EU operates, it was an natural extension of the organisational remit, albeit at this stage remaining very much an inter-governmental activity. Despite the apparent logic inherent in the creation of a European approach to foreign and security policy, there is no guarantee of success, with the unwillingness of several member states to accept the necessity that the EU move in this direction likely to make any negotiations on a CFSP a long and drawn out affair.

That there is a role for the WEU within this process can be accredited to the broader membership of the EU and differences in opinion between the members as to the level to which the EU should become involved in questions of commonality in this area. Although the WEU had accepted the need to develop its membership, there is advantage to be gained in having a separate list of contributors to those other organisations that would serve a similar purpose. Thus it is to the issue of the WEU's developing membership that we now turn.

V. The Growing Membership

As suggested above, the original five signatories of the 1948 Brussels Treaty, were joined in 1954 by the accession of the Federal Republic and Italy. With the addition of Spain and Portugal in 1990, the membership had reached nine and when, most recently, the Greeks became full members, the current ten had been reached. All ten of these states are also members of both the EU and NATO.

The need to involve the remaining European membership of these two organisations came onto the agenda at the June 1992, Petersberg Summit, in Bonn.[31] A means was sought of introducing the eight European members of the EU and NATO, which were not already party to the WEU, into the discussion of defence and security issues at a level which satisfied both them and the existing membership. Thus, two new types of membership were announced for states which were not party to the modified Brussels Treaty of 1954: Observer and Associate. The Petersberg Declaration went on to describe the level of involvement that such alternative status would allow.

The level of Observer member is essentially designed for those states which adopt a neutralist stance on foreign and defence policy but who nonetheless wish to remain seized of developments in this sphere. Thus, active participation in the WEU is largely limited to that which the existing membership invite the Observers to become involved in, with attendance at WEU Council meetings providing the opportunity to remain informed whilst not becoming openly involved.

This contrasts with the Associate members whose role is somewhat different. Whereas the Observers' input is severely limited by desire and by design, the Associates role was created to allow for a much more direct involvement in the activities of the WEU. This tier consists entirely of non-EU, NATO states, all of whom clearly have an involvement in defence and security matters and who are more likely to seek active involvement. With the caveat that they can be excluded if the majority of member states wish them to be, and that they cannot affect the outcome of decisions which require consensus among the other members, the Associate members can speak openly in WEU Council, contribute on an equal basis to military operations and associate themselves with the decisions of the WEU.

As Figure 1. demonstrates, Denmark is the single exception to the rule that states possessing dual EU and NATO membership also have full WEU membership. Instead Denmark participates as an Observer, alongside Ireland, Sweden, Austria and Finland, the four states which share a stance of neutrality and are happy to remain outwith the NATO fold. The three NATO states with Associate status are Turkey, Iceland and Norway.

As the graphic further illustrates, there is a fourth level of WEU membership in addition to the Observers and Associates: that of Associate Partner. While the implications of this are discussed further in Chapter Six, the notion of extending membership privileges to the former Warsaw Pact states and the Baltic Republics was designed to further integrate these states into the emerging security architecture of Europe and followed the relatively successful introduction of the Partnership for Peace concept - although it is to be noted that the WEU has not opened its door nearly as widely as NATO did with the introduction of PfP and the graphic below denotes only those states with some kind of WEU membership.

However, despite the optimism that should surround such an apparently healthy position, with the introduction of new levels of membership a clear signal that the WEU is an organisation which commands respect and attention in the realm of security and defence matters, there remains a persistent belief that the WEU is somehow of less importance than some of its contemporaries. It is seen as yet another obstacle in the way of seamless development of coherent security policies and adds to the bureaucratic minefield that surrounds any debate on European security. This view is heightened by the fact that the membership of the WEU is a clear subset of the organisations it is operating most closely with and to a certain extent can be deemed to replicate that which already exists.

Figure 1. Organisational Memberships

	WEU	*NATO*	*EU*
Austria	Obs.	PfP	X
Belgium	X	X	X
Bulgaria	Ass.P.	PfP	
Czech Republic	Ass.P.	PfP	
Denmark	Obs.	X	X
Estonia	Ass.P.	PfP	
Finland	Obs.	PfP	X
France	X	X	X
Germany	X	X	X
Greece	X	X	X
Hungary	Ass.P.	PfP	
Iceland	Ass.	X	
Italy	X	X	X
Ireland	Obs.		X
Latvia	Ass.P.	PfP	
Lithuania	Ass.P.	PfP	
Luxembourg	X	X	X
the Netherlands	X	X	X
Norway	Ass.	X	
Poland	Ass.P.	PfP	
Portugal	X	X	X
Romania	Ass.P.	PfP	
Slovakia	Ass.P.	PfP	
Slovenia	Ass.P.	PfP	
Spain	X	X	X
Sweden	Obs.	PfP	X
Turkey	Ass.	X	
United Kingdom	X	X	X

X denotes full membership

Obs. = Observer,

Ass. = Associate,

Ass.P. = Associate Partners,

PfP = Partnership for Peace,

As we saw in Figure 1., there is no full member of the WEU that is not also a member of both NATO and the European Union. Of the additional levels of membership those European states that make up the remainder of the memberships of both NATO and the EU provide the second and third tiers of WEU membership through the Observer and Associate blocs. Even the Associate Partner countries are merely a small subset of those states which have signed up with Partnership for Peace.

Yet this seems to be a process which is fully accepted by the states involved and fits comfortably with the overarching vision of a European security structure which is based on the notion of interlocking institutions. With the debate over enlargement of both NATO and the EU moving apace, this fact is likely to become more apparent. While there is little evidence of a symbiotic relationship between the two enlargement processes, there is a core of names most normally associated with a first phase of admission. Thus any discussion of enlargement usually leads us into some discussion of the Visegrad states (Poland, the Czech Republic, Hungary and Slovakia), with Slovenia, the Baltic Republics and the other former Warsaw Pact states forming a second and third tier of possible members. As this process moves forward it would seem likely that the changes in the membership of one organisation will be reflected in a similar, if chronologically different, enlargement process in the others.

The upshot of all of this is that there will be a gradual reduction in any differences between the organisations. Again this is consistent with an era where notions of economic and military security are seen to be more closely linked than ever before.

VI. The WEU's Emergence as an Alternative Forum

It has already been suggested that there are EU members that would rather avoid commitment to the creation of a CFSP. With this opposition likely to bar progress in this field, yet not withstanding the desire to bolster the West European approach to matters of foreign and security policy, the WEU is seen as the ideal instrument for the development of this kind of cooperative action.

By providing a forum for the subset of like minded nations who wish to become involved in such activities, the WEU can be seen to have re-emerged from the shadow in which it had so effectively hidden for thirty years. Whether the WEU eventually forms an autonomous subset within the broader organisation or perhaps even becomes a candidate for submersion within the EU is something that is still in unclear and is discussed further in Chapter Six.

While the Maastricht Treaty and the progress which has been made in relation to CFSP since 1991 give the WEU further credibility, the WEU itself has been deliberately active in seeking to expand its efforts. Building upon the foundation of the Rome Declaration and successive Ministerial meetings, came the Petersberg Declaration of 1992 mentioned above. As well as announcing the two new levels of membership, Petersberg helped clarify the position the WEU hoped to build for itself within European security and laid down the basis upon which the WEU hoped to go about strengthening its operational role and credibility.

Thus the Declaration also included the following key points:

- the members declared themselves willing to make military units available, under WEU authority
- these forces could be used for humanitarian and rescue tasks, peacekeeping tasks and where required combat forces in crisis management and peacemaking tasks.
- announced creation of a WEU Planning Cell to prepare contingency plans and recommend the necessary command, control and communication (C^3) requirements [32]

Also notable within the Declaration was mention of the important role which the experimental Satellite Centre in Torrejon, Spain could serve in terms of operational capacity.[33] It was further declared that the space activities of the WEU would be limited to the full members, at least during the initial evaluation and experimental phase of the Satellite Centre.[34]

VII. Future Development of the WEU

The WEU is often viewed as the poor relation among the clutch of organisations that exist to service the security interests of West European states. In large part this is directly attributable to the remarkably low profile that the organisation has possessed throughout much of its period of existence, with only the most recent past having brought the organisation any discernible level of recognition outwith the politico-military and academic communities. This is a somewhat puzzling situation for an organisation that can trace its lineage back to the Dunkirk Treaty and the first emergence of any form of European collective defence following the war.

While the WEU of the 1990s can be considered the junior partner in the field of European security, dwarfed by the political and military significance of NATO, the WEU is, for the first time since its creation, becoming a valuable, perhaps even crucial, element in the development of the European security landscape. That it took the organisation over thirty years to shrug off the lethargy which had shrouded it since birth is testament to the fact that not only was NATO a clearly superior element in the security of the member states, but also that a raft of additional organisations have coexisted alongside it in the so-called European security architecture. This situation helped to ensure that the WEU had avoided, until recently, efforts to develop its potential as fully as could perhaps otherwise have been expected.

Thus the WEU finds itself competing for attention in a field which contains organisations ranging from the purely Western to the pan-European and from those for whom security and defence matters are their raison d'être to others for whom their primary concerns rest more in the realm of economic and social matters. The shared concern of all these organisations is their reliance on the stability and security of the European continent for the successful pursuit of their individual agendas. Despite the divergent aims and remits of this multitude of organisations it is possible to map a common theme among them all - in the rarefied atmosphere of post Cold War Europe, all, without exception, are involved in an attempt to carve for themselves a niche in the post Cold War European organisational structure. To this end we have not seen the predicted demise of international organisations that can often be considered children of the Cold War, including both NATO and the WEU. Instead we have witnessed the expansion of such bodies with the inclusion of the so-called 'Partner' states of Central and Eastern Europe and a scramble to build an interlocking framework of institutions which it is hoped will provide the stability which was formerly a notable feature of the rigidly bipolar international system.

This move towards interlocking institutions is one of the most significant features of the current organisational climate, with the WEU seen as a crucial component of this emerging pattern. While the fanciful might suggest that the WEU is engaged in mortal combat with NATO in a bid to become the premier security body in the Europe of the next century, this is a misguided interpretation of events. As suggested above, the WEU is not in direct competition with any of these additional bodies, despite overlap in many spheres of activity. As a result, the debate over the future role that the WEU can play in European security concentrates more on how it will be integrated into the existing framework than over how to replace parts thereof. The consequence of this debate is to place the WEU at the very heart of the vision of future European security. Whether this be in conjunction with the European Union and NATO or through some combination of the three, the WEU has become a major player.

If it can be said that in the 1990s the WEU's time has come, then it must also be said that it has been a long time coming, almost fifty years. However, as already indicated, it has not been totally idle throughout that half century. In no small part due to the work of its Assembly, the WEU has developed a long history in specific aspects of European security, not the least of which is the role of space activities. Subsequent chapters will assess the organisation's ongoing interest in this important feature of contemporary security and will look at the potential for the WEU to make this one of the cornerstones of its involvement in the preservation of security in Europe.

Perhaps it is unfortunate that the first period of prolonged exposure the WEU has enjoyed since its inception, has coincided with what is often seen as a demonstration of its ineffectiveness, with the abortive involvement in the early stages of the Bosnian crisis only salvaged by participation alongside NATO vessels in Operation Sharp Guard. The relative size and scope of the organisation has also counted against it being considered an equal to, or even potential replacement for, NATO; with its headquarters and staff representing a far less significant body than that which exists to serve the needs of its more illustrious counterpart.

It is to this operational capacity that we now turn our attention, examining the role that the WEU has sought to carve out for itself as the premier organisation in Europe looking at the role of military space, before considering the potential for this activity to stand at the heart of what the WEU can offer European security.

ENDNOTES

[9] Urwin, D., (1991) *The Community of Europe: A History of European Integration Since 1945*, (Longman, London and New York, 1991), p.9

[10] The modified Brussels Treaty of 1954 renumbered the Articles - with the provision contained in Article IV of the 1948 version becoming Article V in the modified Treaty which is included as Appendix One

[11] Article III, Appendix One

[12] Article IV of the 1948 Treaty

[13] Taken from the Charter of the United Nations, San Francisco, 1949 Chapter VIII., Article 53.2

[14] Germany had been split into four zones of military occupation in the aftermath of World War II, with the French, British and Americans in the west of the country and the Soviet Union controlling the eastern areas. On a smaller scale the same thing had happened to Berlin, although it was physically located well within the Soviet controlled sector of Germany.

[15] Churchill, Sir W., Speech at Westminster College, Fulton, USA, 5 March 1946

[16] By May 1949 the allies had airlifted 2.3 million tons of supplies into the besieged city of West Berlin. Taken from Sheehan, M. & Wyllie, J., The Economist Pocket Guide to Defence, (Blackwell, London, 1986), p.32

[17] See Article 5 of the Washington Treaty which is included as Appendix Two

[18] Archer, C., *Organising Western Europe*, (Edward Arnold, London, 1990) p.18

[19] The ECSC had been designed to integrate the potential heavy armaments sectors of West German industry into an international forum, thus allowing for industrial development, while also enabling rigorous checks to be put in place to ensure this was not being abused

[20] Preamble to the modified Bruusels Treaty, Appendix One

[21] Archer, C., *Organizing Western Europe*, (Edward Arnold, London, 1990), pp.168-175

[22] URL http://www.fco.gov.uk/weu/purpose.html *WEU: Purpose and Activities* taken from the UK FCO World Wide Web Pages 31/7/96

[23] Cahen, A., 'Relaunching Western European Union. Implications for the Atlantic Alliance' *NATO Review,* No. 4, 1986, p.7

[24] Urwin, D., *Dictionary of European History & Politics*, (Longman, London, 1996), p.164

[25] WEU Press and Information Service, *WEU, History, Structures, Prospects*, (Brussels, 1995), pi

[26] Cahen, A., 'The Western European Union (WEU) and NATO: Strengthening the Second Pillar of the Alliance', *The Atlantic Council of the United States Occasional Paper*, (Washington DC 1990), p.19

[27] WEU Council of Ministers, *Rome Declaration*, 1984

[28] WEU Press and Information Service, *WEU, History, Structures, Prospects*, (WEU, Brussels, 1995), p.9

[29] WEU Council of Ministers, *Hague Platform*, October 27, 1987, Part III, Paragraph 2

[30] WEU Press and Information Service, *WEU, History, Structures, Prospects*, (WEU, Brussels, 1995), p.9

[31] See WEU Council of Ministers, Petersberg Declaration, October 23, 1954, included as Appendix Five

[32] Appendix Five, Part II

[33] Appendix Five, Part I, Paragraph 15

[34] Appendix Five, Part III, Space Activities

CHAPTER THREE

DEVELOPING AN INTEREST IN SPACE

I. Introduction

The most strident and persistent proponent of European military space cooperation over the last 30 years has without a doubt been the Assembly of the Western European Union. A deliberative, rather than a decision-making body, the WEU Assembly has, however, always attracted some degree of governmental attention. As the overall influence of the WEU as a whole has increased since the mid-1980s, so too has the influence and importance of the deliberations of its Assembly.

Since the early 1960s, at least 40 reports to the WEU Assembly, from its various committees, have been concerned either principally, or in large part, with European space activities. Significantly, perhaps indicating the growth in importance of space-related topics in both the European political and security arenas, the 1990s have seen several reports pertaining to space-related issues presented to the Assembly for its consideration, with many relating to military space applications such as Earth observation and ballistic missile defence.

By examining the series of reports relating to space which the Assembly has considered, it is possible to trace the growth and development of the WEU Assembly's interest in this area.

II. First Steps into the Analysis of Space Activities

The first report dedicated to space originated from the Committee on Defence Questions and Armaments in May 1962, on 'European Space Organisation'.[35] This report noted the possibility of cooperation between the USA and the USSR in certain aspects of space endeavours, particularly weather and communication satellite programmes, and preparations for the probing of Mars and Venus. It welcomed the efforts which had gone into the draft agreements for the European Space Research Organisation (ESRO) and the European Launcher Development Organisation (ELDO), while at the same time expressing regret at the lack of urgency which had been displayed in the actual, operational establishment of these two bodies.

The report also expressed concern that the establishment of these two agencies placed space outwith the overall framework of future European organisation, in particular commenting on the need for such activities to be under firm political control. The Committee suggested that it would have been better to keep the field of space activity within the overall framework of European integration which was developing at that time, perhaps by the creation of a separate Community, which would facilitate its incorporation into a European Union, should one develop at a later date.

Thus the report recommended that the Council of WEU invite member governments :

1. To bring the whole of the European space research organisation into the framework of a European Political Union, of which the United Kingdom must be a full member, by adapting accordingly the draft treaty establishing the Union;
2. To entrust the final responsibility for space research and development to a committee of Heads of State or of Government;
3. Meanwhile, until this has become a practical possibility to instruct the Council in cooperation with the Prime Ministers to take all steps to speed up and co-ordinate the works of the European Space Research Organisation and the European Launcher Development Organisation.[36]

A number of interesting points emerge from this relatively short report. Firstly, it clearly indicates the importance which the Assembly placed on space activities in the early 1960s. Not only was the point made that clear political control of space activities was important, but they called for the final political responsibility to rest at the highest possible level - with a committee of Heads of State or Government. Space, it was felt, was too important an issue area to be left to lesser ministers. Secondly, and just as significant, the development of space activities was regarded as an integral part of much wider political machinations. It was just one aspect of a more widespread desire to move ever closer towards the political integration of Europe. To this end, the first recommendation of the report called for the establishment of a European Political Union. In this respect, the report was not just an examination of the organisation of European space activity, but a clear statement of desire in the broader arena of European politics.

In fact, the whole report had been prepared as a follow-up to a chapter in an earlier report of the Committee, a Report on European Security, 1956-61.[37] One chapter of this earlier report had dealt with the question 'Space War or Space Cooperation: The European View', which expressed the conviction that:

> '... if the cold war threatened to reach out into space it would constitute a serious threat to world peace, but that peaceful cooperation between the two great powers could benefit all mankind.' [38]

This was a comparatively widespread view in the early part of the 1960s. Manned spaceflight had just become a reality, while the Cuban missile crisis had served as a reminder, if one were needed, of the inherent dangers of the Cold War escalating into a nuclear confrontation. The 'Space Race' between the two superpowers was reaching its height, as both sides sought to win propaganda victories over the other in this new technological arena, and by so doing attempting to prove their superiority to the rest of the world. Of particular note in this report was the view that as long as there was no decisive agreement that space should be used for purely peaceful purposes, it was not sufficient for the US to be the only western competitor of the Soviet

Union. This resulted in Assembly Recommendation 69 in which it was recommended that the WEU Council should promote the joint preparation of a western space programme by the United States and the European countries, combining the technical and financial resources of the European countries and the US.[39]

Little wonder then that the Europeans saw this as a critical area, largely in relation to security. Although the thrust of European interest in space was civilian, rather than military, they could not ignore the possible exploitation of the military dimension by both the USA and the USSR.

This theme was continued in 1964, with a report on 'Progress in space technology and possible repercussions on strategy'.[40] In this, moves under the auspices of the United Nations which sought to ensure the peaceful development of space were noted, but with little optimism that they could be realised. Having compared the European investment in space of some $100 million annually with that of the Americans - $2,600 in 1963, a figure which it was projected would have reached some $5,600 million by 1965, the report concluded that:

> 'Europe seems resigned to stagnating in space ... The use of space for aggressive purposes is a fact, and we must subordinate our behaviour to this fact until such time as a real international agreement on disarmament will invalidate it.' [41]

Consequently, the report's recommendation to Council was as follows:

> 'Considering that the security of Europe and the European countries requires a much greater effort in the field of space ... [The Assembly] Recommends to the Council that the member governments, in their capacity as member governments of ELDO, be urged to call for a new programme for this organisation with the objective of reaching the most advanced level of development as soon as possible [and] that for this purpose the cooperation of the Government of the United States be sought in the framework of an Atlantic partnership.' [42]

This report seems to have struck a chord with the WEU Assembly. Perhaps jolted into action by the fear that the European countries would be left behind in this area of technological development which, at that time, seemed to be of considerable concern both in relation to security, and also to technological development, they decided that space was too significant an area to be dealt with as a comparatively small part of the remit of the Committee on Defence Questions and Armaments. Consequently, in what can be seen as a significant marker of its perceived importance at the time, the Assembly of WEU in November 1964, established the Committee on Space Questions, with responsibility for keeping under review all questions appertaining to space. In October 1965 it produced its first report, 'The State of European space activities: the political choice'.[43]

This report dealt very largely with the non-military aspects of European involvement in space, but in doing so again reiterated the necessity for the governments of the member states to lay down guidelines for long-term space goals, for without such guidelines it was feared Europe would 'stand no chance of keeping abreast with Russia and America in this field.'[44]

Of course the interest of the WEU Assembly in space matters was not happening in total isolation. At this time strenuous efforts were also being made in a number of different fora with reference to European space activity. Consideration was being given to the most appropriate international, institutional setting for the furtherance of European space interests, and several European international organisations with elements of responsibility for such a European strategy were evolving.

II. The WEU and European Space Organisations

Indeed, by this stage, the two principal international organisations for European cooperation in space research and development had come into being, ELDO in February 1964 and ESRO in March 1964. The first report of the Committee on Space Questions outlined the purposes of these organisations, in essence to facilitate cooperation between member states, and also between the member states and the wider international space community, and pointed out that, even at this very early stage in the life of these organisations, ESRO seemed to be on a firmer footing than its counterpart, ELDO - a somewhat prescient observation, as events would later prove.

The report went on to examine Europe's future in space, in terms of both scientific and launcher research and again returned to the question of cooperation with the United States, pointing out that 'because of the heavy financial investments involved, it becomes of paramount importance that the efforts are not duplicated, or dissipated unnecessarily'.[45] The particular emphasis here was that 'Europe must seriously question the advisability of attempting to build first generation launchers at a time when the United States of America is so well advanced in the technique.[46]

The report identifies three possible courses of action for Europe:

- continue the present limited and uncoordinated programmes with their lack of any real objectives
- decide to contract out of space research and development altogether
- to so design the infrastructure of the European effort that the possibility of joining with the United States of America in a common assault upon the vast field of space research problems becomes a reality [47]

Without doubt, this report clearly met its objective of attempting to focus the Assembly's attention on the European space programme, and to identify the major political decisions which had to be made at that time. In outlining the alternatives above, it succinctly identified the key issues facing Europe in the mid-1960s in relation to space activity in general. The principal recommendation resulting from this was simple:

> '...that the member governments should prepare a ministerial conference so that decisions may be taken on European activities in space and the means of carrying them out, as well as on the organisation for implementing such decisions.' [48]

The relatively gentle reservations contained in the first report from the Committee on Space Questions, regarding ELDO and its operations were more than justifiable. Thus, while perhaps disappointing, it was not altogether surprising that the title of their next report, just seven months later was 'State of European space activities - The ELDO crisis'.[49]

Referring to the conclusion contained in an earlier report to the Assembly, that it was better for Europe to have one overall space organisation rather than separate national organisations set up under special treaties,[50] this report highlighted the fact that, notwithstanding this suggestion, the governments had thought it better to start with European cooperation by special treaty, but from the beginning it was clear that ELDO lacked the impetus of political direction.[51]

Essentially, the 'ELDO crisis' which had quickly emerged was that the French, in 1965, proposed that the initial ELDO programme be changed, halting all work on the originally planned ELDO A launcher, and instead re-focus the programme to start building the ELDO B launcher straightaway. This however, was not technically feasible, and the original programme continued as planned. As the report to the WEU Assembly indicates:

> '... this incident shows the difficulty of pursuing such a project when it is not under the direction of a political authority. The ELDO Council is composed of civil servants from various administration of the member countries and has a large number of alternate delegates. In the above-mentioned instance, the French proposal was made without advice from the technical staff.' [52]

However, this was only the first element of the ELDO crisis. The second, and potentially much more grave element came from the other side of the English Channel. In February 1966, the British government in London had expressed grave concern over the future of ELDO, in relation to both the technical and financial aspects of the organisation. It was feared that delays in the completion of the ELDO A launcher meant it would only become operational at the end of 1968 or the beginning of 1969, by which time it was feared it could well have become obsolete. Britain was also extremely concerned about the cost escalation of the programme, with the British contribution running at almost 39 per cent compared with the significantly smaller contributions of the French (24 per cent) and the Germans (22 per cent).[53] The fear which such sentiments engendered was that the UK may be considering withdrawal from ELDO, despite it originally having been set up largely at the behest of the British, who had been keen to internationalise their Blue Streak rocket programme: something which had proved too costly to sustain at the national level.

In a sense it is difficult to overstate the potential gravity of the situation. At the time there was a very strong feeling that if the situation were not somehow resolved, the nascent European space programme might well be doomed, virtually from the outset.

At the time this WEU Assembly report was being considered, the relevant national ministers were still reviewing the ELDO programme. With this in mind, the Assembly, stating that they were 'convinced that if Europe were to discontinue a launcher development programme this would have irreparable consequences for the future of scientific and industrial development in the countries of Western Europe, and considering that monetary financial difficulties which might be encountered by a country associated with the ELDO programme should not constitute an insuperable obstacle to continuing this programme', recommended that every effort should be made 'to overcome the difficulties in European cooperation in the field of space-vehicle launchers.'[54]

The reference to national financial difficulties was a clear reference to the UK position, representing as strong a statement as could be made diplomatically about the internal considerations of a member state and their potential ramifications for European space cooperation. As it was to turn out, however, as will be discussed later in this chapter, national considerations, as is so often the case, were to prove to be the crucial factor.

This situation does, however, offer a clear example of the potential role of the WEU Assembly in the European space effort. Although not formally charged with any specific role or responsibility, it provided a relatively neutral international arena within which consideration of such questions could take place. It is true to say that the Assembly has consistently served as one of the most 'European' of international fora. The so called 'national interest', all too often of paramount importance in other international organisations, has always been seen to take less of a prominent role in the discussion of the

Assembly. This is not to say that the members are not mindful of their role as national parliamentarians, but they do seem to be capable of leaving their national prejudices behind and adhering to the discussion agenda proffered by the organisation and not the one which their national interest might promote as preferable.

At the same meeting of the WEU Assembly, another report was put forward by the Committee on Space Questions. Entitled 'Global communications by satellite - Europe's role',[55] this report, cognisant of the likelihood of a global international tele-communications regime emerging in the near future, was concerned with the lack of a coherent European position and called for a detailed study of problems which would arise in the field of telecommunications, and pressed for the commencement of immediate preparations for an agreement with the United States with regard to a world-wide satellite communications system.[56]

1966 was proving to be a very busy year for the Committee on Space Questions. Having presented the two reports outlined above to the first part of the Twelfth session, they submitted another two reports to the second part of the session in November of that year. 'Juridical Problems and space policy',[57] was a generally comprehensive examination of a wide range of aspects of international law relating to outer space, and found its context against the background of negotiations on what would become the Outer Space Treaty.

However, of much greater significance was the Committee's fourth report of the year, 'State of European space activities'.[58] This continued the theme of previous, similarly entitled, reports and represented a further examination of Europe's space policy at the time. As the introduction to the report states:

> 'The Assembly's continuing concern, as expressed in the various recommendations it has adopted, is to give the European space effort the continuity and political impetus which is implied in a long-term task requiring ever larger technical, intellectual and financial resources.' [59]

Overall, this is a much more upbeat report than some of its predecessors. At this time, November 1966, it was felt that progress had been made on a number of problem areas, not the least of which was the situation regarding ELDO. A Ministerial conference of ELDO states in July 1966 had come to an agreement which seemed to keep all the members happy. They had agreed to continue with the original work programme, but had capped the total expenditure, and had also amended the contributions to be paid by individual member states, raising France's share to 25 per cent, Germany's to 27 per cent, and reducing the British contribution to 27 per cent.[60] This, it was felt, was a fairer distribution of costs, and one which would ensure the UK remained content to be an active participant.

In addition, the ELDO Ministerial conference had also proposed that 'ESRO and CETS (the European Conference on Satellite Communications) agree to the creation of a Co-ordinating Committee, charged with the immediate co-ordination of the functioning of the three bodies.'[61] This was intended to ensure greater congruence between the work of the three different bodies directly concerned with European space activity. This idea was grasped firmly in the report to the WEU Assembly, which called for the work of this Co-ordinating Committee to be accelerated, thereby laying the foundations for a European space community based on the merger of the existing organisations, ELDO, ESRO, and CETS. In making this call, they were essentially re-iterating the position which they had held for several years, that the only really effective way to encourage a truly European space endeavour was for it to be governed by a single international body.

At this time, the Committee structure of the WEU Assembly itself was again re-organised, and the areas of competence previously under the auspices of the Committee on Space Questions became part of the remit of the new Committee on Scientific, Technological and Aerospace Questions. While this could be seen as a means of downgrading the importance of space issues - removing it from the sole concern of one committee, to one aspect of a committee with a much broader remit - it can in fact be seen to have achieved the reverse. Thus, by placing the examination of space issues alongside the more mainstream and generalised issues of science and technology, the space agenda of the WEU Assembly gained an even greater level of legitimacy. In any event, whatever the motivation behind the restructuring, the net result was no less consideration of space issues within the newly configured committee system.

Their initial report on space matters was produced in May 1967, and was a return to one of the other major concerns that had been expressed over the years - communication satellites. Thus 'State of European space activities - European space and satellite technology and international competition',[62] returned to the concerns of international developments in the field of communication satellites, and in particular the steps which should be taken to ensure that Europe did not lag behind other countries, especially the United States, in this area. It noted the COMSAT Act, which the US Congress had passed in 1962, with the aim of setting up a body for the commercial exploitation of satellite communications, and to intensify US policy in this field.

The report examined COMSAT's policy in business and in placing orders, and concluded, not surprisingly, that the way for Europe to proceed was to establish a comparable organisation. It therefore proposed the establishment of a body, tentatively named EUROSAT, with capital from both the public and private sectors, with shares held by national holding companies in each of the member states. EUROSAT would be tasked with developing and installing national and regional television and communication satellite systems and, in collaboration with CETS, organising the European contribution to INTELSAT.

III. Co-ordinating European Space Activitity

The second part of the Assembly session, in November 1967 received a further general report on the 'State of European space activities' which returned to the theme of the overall organisation of Europe's activities in space. This report detailed current space activity in all the member states, and also reported on the establishment of a European Space Conference (ESC) which had been convened, at ministerial level, to consider a co-ordinated space policy. The first ESC was held in Paris in December 1966 at which time a complete list of all European national or international space programmes which were underway, or envisaged, was drawn up. The ESC also deliberated upon Europe's requirements in the space field. The second ESC was held in Rome in July 1967, by which time it had been joined by all European states which were participating in any of the three existing bodies. Significantly at this conference, it was agreed that the ESC should be institutionalised and meet on a more regular basis. In addition to this it was also agreed that its decisions would 'have an executive character as far as the governments which adopted them were concerned.'[63]

These developments can be seen as a vindication of the position which had been held by the Assembly of WEU for some time - namely that European space policy was of such importance that it should be co-ordinated at ministerial level, across the countries of Europe. The ESC was a major step in this direction, providing a formal institutionalised body concerned primarily with the coordination of European space activity, and within which discussion could take place at ministerial level. It was not however the perfect solution, nor was it to be the final one. Once again the November 1967 report to the WEU Assembly, not content with the progress that had been made in this direction, raised a new series of questions regarding European space cooperation. In the light of a resolution adopted in October 1967 by the Council of Ministers of the European Communities relating to technological cooperation, especially in the field of data processing and telecommunications, this report raised the question of the relationship between space activities and the European Communities.

It suggested that a political choice had to be made and advised the selection of one of four possible options for the future of the four space related bodies that now existed in Europe: ESC, ELDO, ESRO, and CETS.

These options were:

- incorporating these bodies into the European Communities
- associating them with the Communities
- leaving them independent of the Communities as at present and therefore open to different membership
- amalgamating them in a new framework [64]

That such questions were being raised within the Assembly shows a considerable degree of foresight on its part. The whole question of the relationship between the European Community and the European space industry was to command much attention in the 1980s and 1990s, and is still something which is under debate today, a point which will be discussed further in Chapter Eight.

In May 1968, the Committee on Scientific, Technological and Aerospace Questions submitted a report to the Assembly on the 'Soviet orbital bombardment system'.[65] This followed the announcement in November 1967 by the US Secretary of Defense that the Soviet Union had developed a fractional orbital bombardment system (FOBS). The report concerns itself largely with the relationship between such a system and the Treaty on the Principles Governing the Activities of States in the Exploration and Use of Outer Space (the Outer Space Treaty) which was adopted by the United Nations General Assembly in December 1966. It also points out that the stated intention of the US to develop a new weapon to counter the FOBS represents and increased emphasis on the military space programme of the US.[66]

Of rather greater significance for this study, however, are the conclusions that this report reaches in terms of the impact of such developments on Europe. It raises concerns that the lack of intermediate range nuclear missiles in Europe, and the existence of an anti-ballistic missile system in the Soviet Union had weakened the Western European military position. In order to give a feeling of the thinking of the time, of the fear which such moves engendered in the Cold War atmosphere of the 1960s, it is worth quoting one paragraph from the report in full:

> 'With the combination of the Soviet missile threat and the FOBS, plus the fact that Western Europe has neither an anti-ballistic missile system nor an adequate radar system, little can be done in the way of defence. The Russians, moreover, must have their reasons for creating this new system and the western world cannot overlook its yet unknown potential. Involvement in the complex system of space technology has far-reaching consequences for our life and our survival. The Soviet

Union might very well unveil further space capabilities which will not be to the advantage of the western world and especially Western Europe. If our continent is not willing to organise and pay for its space research and development, we shall never have any insurance against military surprise. If we do not make a greater effort in space technology, the outlook will be even more bleak in the years to come.' 67

This clearly indicates the level of concern which moves by the Soviet Union generated in Western Europe. This is, of course, unsurprising in the Cold War atmosphere of the time. Throughout the 1960s, the all pervasive fear was of Soviet expansionism, resulting in conflict within the European theatre. Thus, almost all consideration of European security was undertaken within a frame of reference which highlighted the Soviet threat. However, of more importance in assessing the role of the WEU Assembly and the development of European space policy, this helps to highlight the dawning realisation that space activity was no longer simply another area of international competition or collaboration, rather it had very much been placed squarely on the international security agenda. Up to this point in time, most consideration in this field had been devoted to the organisational and technical aspects of space activity. This report, however, is the strongest indication since the 1961 report on European security which had considered the question of space war or space cooperation that the Assembly was still mindful of the security implications of space. In this respect it was, for many years, virtually the only European body to consider such security implications. Indeed, by the 1980s, this was its primary concern with space.

The following year saw a return to another familiar topic of consideration, with the submission of a report to the Assembly on the 'State of European space activities - INTELSAT',[68] which largely reiterated previous arguments concerning the European situation with regard to satellite communications vis-à-vis the American situation, and called on the European countries to 'concert their positions before the final discussion in 1970 on the future INTELSAT agreements.'[69]

In a sense, this led on to a report in 1971, which was potentially one of the most significant reports on a space issue yet produced by the Committee on Scientific, Technological and Aerospace Questions, which drew together a number of strands which had appeared in previous reports. 'Prospects of space collaboration between the United States and Europe'[70] covered a wide range of aspects of potential collaboration. It was written against the backdrop of the US post-Apollo programme planning, which at that time was still under active consideration, since no final decision on the Space Shuttle programme had yet been taken. One of the strands of the US post-Apollo programme, however, was international participation and cooperation. To this end, discussion had been ongoing between the Americans and the Europeans since 1970 on the possibility of the Europeans financing and building part of the space transportation system, specifically a space tug, which would be taken into orbit by the Shuttle.[71]

The thrust of this report to the WEU Assembly therefore was to examine the prospects of scientific collaboration, including unmanned probes to other planets in the solar system, and technological collaboration in the Skylab and post-Apollo programmes. It further addressed more general issues such as access to US launch capabilities. The report was of the opinion that American-European collaboration would be advantageous to both partners, and that the Americans were keen on the prospect:

> 'During the Committee's visit [to the US] it became quite clear that the American government considers collaboration with Europe in the post-Apollo programme and thereafter as a serious objective. It wants to secure this collaboration and it was for this reason that on 1st September the United States Under-Secretary of State ... wrote to the Chairman of the European Space Conference ... stressing that the United States certainly did not wish to stop European progress in the space field.'[72]

Accordingly, the report recommended that: all member states should re-examine, with the United States, the overall content and purposes of the space programmes of the late 1970s and 1980s; WEU Ministers for Science and

Technology should hold a conference in order to co-ordinate their polices and prepare priority decisions; and that member governments should set aside the funds necessary to ensure the application of these decisions. The final recommendation was that the member governments should 'implement the decision on the establishment of a single space organisation, NASA's European partner, capable of negotiating, establishing and developing the joint programmes.'[73]

With the benefit of hindsight, of course, it is easy to dismiss such an enthusiastic embracing of transatlantic cooperation in space activities as naïve. Outwith the forum of the WEU, considerable suspicions had emerged concerning US-European cooperation. These concerns remained to the fore, and thus were to severely limit such cooperation.[74]

The recommendation, however, that a single, unified European space organisation should be established, was of course to meet with rather more success. It is perhaps ironic that the report on possible US-European collaboration should largely come to naught, while the recommendation therein on the unifying of the European space scene, was the element which was to be more actively pursued. Again, the WEU Assembly had pursued, with single-minded determination, an organisational aspect of European space activity, with a foresight unparalleled in any other international forum of the time. Repeated calls for a unified, coherent and integrated European approach to space issues were ultimately to be rewarded by the creation of the European Space Agency.

The Committee returned to a much more directly military theme in 1972 with a report on 'Orbital stations and the civil and military consequences'.[75] This was inspired in large part by the US Presidential decision in January 1972 that the United States should proceed with the development of the Space Shuttle which would undoubtedly result in the existence of manned and unmanned space stations. The Committee on Scientific, Technological and Aerospace Questions was therefore concerned with the capabilities for civil and military control which such facilities would offer. 'Like the large unmanned geostationary satellites, these space stations therefore provide additional means of controlling our planet from space, both politically and militarily'.[76] The report set out to examine a number of aspects of such developments, including juridical aspects, the US and Soviet programmes and the possibilities for Europe. It also highlighted the military nature of much space activity, commenting that 'A large number of military support activities in space are passive in character and widely practised by the United States and the Soviet Union'.[77] It went on to call for the establishment of a space study group whose tasks would be:

- to analyse Europe's civil and military space potential and evaluate possible threats to its political, economic and cultural development; to study trends in space concepts and operations and compile forecasts of space technology;
- to assess the sums required to preserve and extend European space research, technological development and operational capabilities;
- to study the implications of space in possible arms limitation agreements and pursue the policy which was started with the ban on nuclear tests in the atmosphere and on the use of outer space for military purposes [78]

The significance of this proposal is clear. For the first time, there were calls for a mechanism to specifically allow for the examination of the military potential and implications of space on a European basis. Although it would be several years before the Assembly next received a report solely concerned with the military aspects of space, it does again indicate that this dimension was always in mind at the WEU Assembly. At this stage, it is worth reiterating the key difference between the WEU and other international organisations. The WEU is the principal European body with specific responsibility for the discussion of security matters. Thus, their concern with space efforts has always had this as the underlying context of such discourse. This makes the WEU different from the plethora of other European organisations which have an interest, but not a specific one, in this area. The years that followed have served to reinforce this point again and again.

In the interim, there was a further significant development in European space policy, which resulted in the Committee on Scientific, Technological and Aerospace Questions bringing forward, in May 1974, in reply to the 19th

Annual Report of the WEU Council a report on the European Space Agency.[79]

Finally, perhaps at least in part as a result of the consistent and repeated calls for such a move by the Assembly of WEU, a process was set in motion designed to establish a single unified organisation to be responsible for European space activity - the proposed European Space Agency (ESA). In fact, it was to be another year before the ESA Convention was finally signed, in May 1976, with the organisation not technically entering into force until October 1980, having taken four years for all the states to ratify it. Despite this time lag however, the ESA effectively commenced operations in April 1975.

The Committee report, noting the progress which had been made towards this Convention, and the decisions of the ESC in 1973 as to the work programme of the proposed new agency, did however contain a number of reservations. They had consistently recommended that the best route for European space cooperation was not only through a unified agency, but one that was within the overall framework of the European communities. This was not to be the case.

> 'For political and military reasons the governments have again decided to establish an organisation for limited specified purposes; limited in the sense that the new organisation will deal only with space activities for peaceful purposes and therefore cannot be concerned with military aspects. Several national military space activities exist, for instance for communications, meteorology and reconnaissance, and the development of launchers often touches on developments in the field of military missiles. The governments have also limited the activities of the organisation by spelling out in the convention the projects to be handled. They did not want a Community space programme to fall within the purview of the Commission of the European Communities.'[80]

The report went on to point out that since it did not seem possible to establish a Community space programme, the member governments of the agency would have to deal with all problems on a multilateral basis, which past experience with ESRO and ELDO had shown could be done, although it was not easy.

The establishment of ESA did, despite the weaknesses suggested in the Committee report, represent a watershed in the development of European space policy, albeit on the civilian side. It not only provided the unified, single agency approach for running specifically designated European projects, but it also incorporated the notion of the 'Europeanisation' of national space activities. Not only did the Convention, as finally adopted, lay stress on positive internationalism, within the European context, it obliged member states to offer cooperation within the overall ESA framework on space projects which they initiated. The European dimension was further secured by the provisions of Article VIII of the Convention which effectively 'forced' member states to use European launchers whenever this was possible.[81] The coming of ESA can therefore be seen as a very positive, indeed forceful, step in European collaboration in space activities; something which the WEU Assembly had over many years been calling for.

However, it did only relate to civilian, peaceful space activities. As indicated above, from the outset the WEU Assembly regarded this as a drawback. There was still no formal, institutional forum which was concerned with the military aspects of space, and European security. If anything, this perceived omission strengthened the resolve of the WEU to concern itself with this aspect of space activities. Although, at this point in time, the WEU as an international organisation was almost dormant in terms of its impact, this attitude towards the military significance of space was to become an important factor after its revitalisation in the mid-1980s.

The reports which the Committee on Scientific, Technological and Aerospace Questions produced in the period between the mid-1970s and the mid-1980s essentially followed the pattern of topics with which they had been concerned with up to this time. November 1975 saw a further report on 'United States - European cooperation in advanced technology'[82] which examined various approaches to international cooperation, namely the Community approach, the Atlantic Alliance approach and the OECD approach, and then went on to examine cooperation in specific fields, one of which was space.

This was a theme picked up again in May 1978 with a further report on 'United States - European cooperation and competition in advanced technology'.[83] This, rather more detailed report, followed a visit by the Committee to the Untied States in April of 1978. This report, amongst other areas, studied cooperation (and competition) in communications and aerospace was of note in that for the first time it recommended that cooperation in respect of advanced technology projects should include military satellites and other military communications equipment whenever possible.[84] Again, this is indicative of the way in which the WEU, alone amongst international organisations, sought to bear in mind the military, or security aspects of international space activity. This was brought home even more vividly in two other reports from the Committee on Scientific, Technological and Aerospace Questions, also submitted to the Assembly in 1978, both of which concentrated on 'Application satellites'.[85]

The first of these reports largely focused on the programmes of the ESA including: the Orbital Test Satellite programme; the overall communications satellite programme; Meteosat, the meteorological satellite programme; and the earth resources programme. In referring to these ESA programmes, the report calls for a number of specific measures which would ensure a substantive European presence in the exploitation of space (including a decision to proceed with a first series of Ariane launchers).

However, once again the specific interest of the WEU was evidenced by a specific section of the report dedicated to the topic of military application satellites. Examining areas including the Navstar global positioning system, communications, geodetic early warning and meteorological satellites, the report specifically noted 'that, although some member countries support their military forces, or realise the need to do so through military applications satellites, there are no plans for a proper European programme to terminate Europe's total dependence on United States satellites in spite of the fact that many European satellites are, technically speaking, on a par with American satellites'.[86]

This is a significant statement. While it may seem merely to reiterate the point made many times before, concerning the lack of a concerted European policy in relation to military satellites, the language used is, in itself, indicative of an emerging European determination. The reference to terminating 'Europe's total dependence on US satellites' is far and away the clearest reference yet to a feeling that such dependence on Europe's transatlantic security partner may not be an entirely healthy state of affairs. This was a theme which was destined to reappear over the next decade, with potentially far-reaching consequences for the very nature of European security arrangements by the 1990s.

The second report on 'Application satellites'[87] was submitted to the Assembly in October 1978. This followed the pattern of the first part of the report, again with specific reference to military satellite systems, however, on this occasion the recommendations related only to ESA and other civilian aspects.

This focus on satellites was continued in 1980 with a specialised report on the topic of 'A European earth resources detection satellite programme'.[88] This highlighted the infrastructure which already existed in the sphere through the ESA Earthnet and Meteosat programmes, and was of the opinion that then current developments within Europe through the Spacelab and French SPOT programmes should be regarded as the foundation for further European collaborative endeavours in relation to earth observation. Once again, however, not content to consider the civilian implications of such technology, the report specifically addressed the issue of the military implications as well:

> 'Mindful of the military surveillance and reconnaissance implications which in addition to the long-term commercial desirability of a European remote-sensing satellite programme enhance its strategic importance to European nations, the Committee recommended that the Council 'encourage within the Independent European Programme Group the concerted study of the military requirements for remote-sensing satellites on a European basis, the definition of any resulting satellite projects and their economic and efficient procurement'.[89]

Again, this is significant. For the first time a WEU report called for consideration of the requirement of a European remote-sensing satellite system for military purposes. As will be shown in the following chapter, this was to become a recurrent theme of the WEU Assembly, which would be regularly revisited over the ensuing years, and ultimately result in the establishment of a new WEU organ in the form of the satellite centre.

The following year, in the Assembly's reply to the Ministerial Council's annual report 883,[90] the military space dimension again drew attention. In an assessment of current US and Soviet military space activities, the Committee on Scientific, Technological and Aerospace Questions noted that 'the United States was becoming increasingly dependent on space systems for the effective use of its military forces'. In addition, it commented that 'the Soviet space programme continues to be more military in character than that of the United States and appeared to be improving in operational capability as well as reliability'.[91] However, the attached recommendation was significant in calling for the exploitation of European military communication and observation satellites, and the investigation of the military implications of space technology, and in its advocation that this should be promoted within the context of the North Atlantic Alliance.[92] In doing so, it seems clear that for diplomatic, as well as strictly pragmatic reasons, the Assembly was endeavouring to ensure that, while concerned about US dominance of the military space element of the Alliance, it wished to develop European capabilities in this sphere in conjunction with the Americans, not necessarily in competition with them.

The comparatively large number of committee reports which the WEU Assembly has considered over the years provides a clear indication of the importance which this body has attached to this policy area. No other European international organisation has exposed space activities to such close scrutiny (with the obvious exceptions of purely space oriented organisations, such as ELDO, ESRO and ESA).

There have been three main themes which have consistently permeated the Assembly's discussion of European involvement in space: the security implications; European cooperation; and Europe's role in more general international space issues, such as telecommunications.

What must be remembered, is that throughout much of the period discussed in this chapter, the WEU was effectively dormant. With its renaissance in 1984 we can recognise a major turning point in its attitude towards such European space deliberations, firmly focusing efforts on the military applications associated with space.

IV. Revitalisation and the Consideration of Space

The mid-1980s was a turning point for the Western European Union. As was outlined in Chapter Two, the organisation began to re-emerge as a key player in the security considerations of Western Europe. Not only was the organisation poised to take a significant role in the overall architecture of European security, but from this time onward its deliberations on space were to focus to a much greater extent on the military aspects of space endeavour.

In May 1984 the Committee on Scientific, Technological and Aerospace Questions submitted a report to the WEU Assembly on 'The Military Use of Space'.[93] This report reflected on the consistent interest which the organisation had shown in the strategic and industrial implications of the space capabilities of its member countries. It emphasised the particularly appropriate nature of the WEU in the field of military space policy:

> 'The Assembly of Western European Union is the ideal forum for informed discussion and debate about the military uses of space. As the sole European parliamentary assembly with specific competence by treaty in the defence field, its statutory terms of reference have lent themselves ideally to serious treatment by WEU of both the strategic and industrial aspects of a European space programme and of the implications for Western European defence of developments in space technology'[94]

Furthermore, the report set out quite clearly the organisational rationale for such a claim:

> 'The organs of WEU lend themselves naturally to the promotion of a Western European strategy and policy for the defensive applications of

space technology. First the Ministerial Council should be used for politico-military consultations at the highest governmental level to bring about a strategic concentration of policy and planning. Second, the Assembly of WEU should be the forum for reflection certainly but much more - the forum for mobilising political commitment to an active European space programme. Third, the Standing Armaments Committee can undertake the detailed technical studies needed and guide the ministerial Council and parliamentary assembly on the implications of the highly complicated technical developments in the space field. Fourth and last, the Agency for the Control of Armaments can examine the arms control and limitation aspects of military space systems, the difficult problems of verification of international agreements in this field and related confidence building measures. In short, WEU is in most respects the only institutional framework for the political control of initiatives and developments in the military uses of space.' [95]

This explanation of the WEU role in relation to the military use of space, is a fairly significant case of an organisation endeavouring to map out a clear role for itself in the overall architecture of the new European security system. Indeed, a similar trait was displayed by NATO at the end of the Cold War as it also sought a rationale for its continued existence. It must however be said that there is a certain logic to the WEU case which belies this opportunistic interpretation of the WEU's attitude towards space activities. It is certainly true that no other international organisation had so persistently considered the prospects for Europe in space in both the civilian and military contexts, with the plethora of reports which were produced by the Assembly during the 1960s and 1970s regarding space, providing evidence enough of a certain commitment in this sphere. It is also true that the particular organisational framework of the WEU did indeed lend itself to it being an appropriate institution, capable of serious study of this field. Thus it was a perfectly logical area of study for the WEU, and in particularly its Assembly and the relevant Committees, to specialise in. It is not placing too great an emphasis on it therefore to suggest that this two part report on military space can be regarded as a turning point in the consideration of European military space policy by the WEU Assembly.

The report outlined the challenge which was posed by Soviet space activity and its inherently military objectives. 'The Soviets have made much progress toward their dual military objectives of global support to military forces and denial of enemy employment of space during wartime',[96] stated the report. Following an assessment of the Soviet position, and a section on aspects of international space law, the report went on to examine current initiatives and developments in defensive space systems. It referred to the points made in an earlier report,[97] which had pointed out that the United States was becoming increasingly dependent on space systems for the effective use of its military forces. It had outlined the four principal elements of the US space development programme as being: improved space surveillance; increased satellite system survivability; development of anti-satellite capabilities; and command, control and communications (C^3) support. The current document also reiterated that the earlier report had, in addition to these points, signalled that major operational capabilities had been achieved by the Soviets in space communications, meteorology, navigation surveys and manned orbital flights - all of which proved a very high level of technical capacity in the Soviet space programme. Finally, again with reference to the earlier report, it was pointed out that in the recommendation attached to that report, the Assembly had called for the promotion of European military communications and observation satellites and the investigation of the military implications of space technology.[98]

This was of particular relevance, at this time. In February 1984, President Mitterand of France suggesting that there should be European cooperation in space-based military reconnaissance or observation programmes which, in time, could lead to European development of a manned military space station had said:

> 'If Europe were able to launch its own manned space station allowing it to observe, transmit and consequently avert all possible threats, it would have taken a big step towards its own defence. To my mind, a

European space community would be the response best adapted to the military realities of tomorrow.'[99]

The Committee's report on the military use of space also, with a high degree of foresight, touched on developments in anti-ballistic missile systems. Such systems had become topical the previous year, with US President Reagan's famous 'Star Wars' speech. However the report suggested that, although European countries had not, so far, been engaged in ballistic missile defence research at any level, certain countries, notably the UK, France and West Germany, may have the technical knowledge to contribute to such a programme. The foresight lay in the fact that this particular topic was to gain considerable attention in Western Europe over the next decade, and indeed, by the 1990s had managed to work its way on to the political-security agenda, something which will be discussed further in Chapter Eight.

Given the wide-ranging focus of the report, the recommendations attached to it were also relatively wide-ranging, but, perhaps shrewdly from a political or diplomatic perspective, none were terribly radical.

Recommendations included:

- a larger European industrial involvement both in NATO telecommunications satellite programmes and in NATO military satellite programmes as well as in the associated ground station infrastructure, in addition to supporting successful national military communications satellites like Skynet (the UK system);
- the commissioning of a detailed analysis by the Standing Armaments Committee of the implications for European defence of developments in military space technology and in particular of Soviet research and development in this field;
- proposing a European surveillance and reconnaissance programme adapting and refining the sensor technologies in the existing CNES Spot project and the ESA ERS-1 project;
- and the setting of clear European space policy objectives and priorities in the course of its politico-military consultation in the key strategic fields of launchers, manned modules, space station

integration, telecommunications, meteorological and remote-sensing satellites and manned reusable service and space transport vehicles

A second part of the report on 'The Military Use of Space' [100] was produced in November 1984, following a visit to the USA by the Committee on Scientific, Technological and Aerospace Questions in July of that year. The rationale for the report is summed up in its first paragraph:

> 'This report will be more descriptive than prescriptive. It will seek to analyse current developments in the military application of space technology in the United States and their potential strategic implications for the Western Alliance as a whole and for Western European strategy in particular. The purpose of this report is to stimulate interest and debate rather than to lay down dogmatically firm proposals.' [101]

Indeed, much of the report was largely descriptive of the United States' position regarding Strategic defence and its implications for arms control - with an appendix containing the US Department of Defense paper on 'Defence against ballistic missiles - an assessment of technologies and policy implications, and the US space station programme. Nevertheless, the Committee recommendations are of considerable significance in relation to the future direction which the WEU should take in the military space field. If the first part of the report on the military use of space, presented in May of that year was a turning point in the focus of the WEU and space, the recommendations contained in the second part were a clear indication of their determination to be the principal international organisation in Europe in this field.

Included in its recommendations were proposals that the WEU Ministerial Council should:

- act as the primary political instrument for intergovernmental concentration on a unified Western European policy towards the military use of space;
- maintain the closest liaison with the United States Government to prevent divergences of view between the American and Western European partners of the Atlantic Alliance about the benefits of the strategic defence initiative, and suggest a space planning group be established within NATO to that end;
- support industrial, technological and strategic reasons for an expanded European space programme and promote enhanced dialogue on related policies and objectives both within the European Space Agency and national governments;
- provide a clear lead and direction to parliamentary and public opinion in favour of a major European effort to meet the challenge of the space age in the fullest sense through increased scientific space experimentation, commercial applications and security-enhancing space developments

Describing this document as 'descriptive rather than prescriptive', in the light of such recommendations, could perhaps be seen as sugaring the pill. There seems little doubt that, taken together, these two reports on the military use of space, were truly the point at which the Assembly of WEU firmly announced to the international security community that in matters related to the military use of space in Europe, and any consideration of a European military space policy, the WEU was the pre-eminent organisation. Strong words for an international organisation which had lain dormant for many years, and was just beginning to stir from its slumbers. Nonetheless, it can also be seen as a very shrewd political move by the organisation as a whole. It was largely true that it had been the only political organisation to devote any real consideration to European space activity over the preceding 20 years or so. Thus, by seeking to make military space its domain, it was not treading on the toes of any other international organisation in the European arena, and was therefore unlikely to cause great offence. Had it chosen an area which was already regarded the domain of another organisation and it had come to a battle between them, there is little doubt the resurgent WEU would have come off second best to either NATO or the European Communities. In largely focusing on military space activity, the other organisations were content to allow the WEU to seek responsibility for a task which, at that time, none of the other organisations had either the inclination or the qualifications to assume. As the ensuing years were to demonstrate, whether through prescience, luck, or sheer dogged persistence, the WEU had picked an area of specialist competence which was to work its way up the international political agenda.

In re-affirmation of this, a colloquy was held in September 1985 on 'The Space Challenge to Europe'. This colloquy, comprising as it did government representatives, parliamentarians, industrialists, and scientists, was intended to provide significant expert input in the preparation of the political questions the WEU Assembly was to consider in December of that year. A number of the points raised in the colloquy are worthy of specific note.

Summing up the non-political aspects of the colloquy, Roy Gibson highlighted a number of areas of civil space usage which were very encouraging. These included the success of the Ariane launcher, the growth in national interest in space throughout the countries of Europe, and the fact that the ESA Council had, at Ministerial level, approved a significant extension of the European civil space programme.

However, significant among the areas which had provided cause for discontent at the colloquy was the fact that the total European investment in space was estimated at one tenth the level of that spent in the United States. It was observed that:

> '... above all the reason most cited for dissatisfaction was the absence of a European military space programme ... the success of European aerospace industry will always be limited until there is a military space investment.' [102]

Christian Lenzer, the General Rapporteur for political questions, indicated that:

'A question of great importance was whether Western Europe as such should develop space activities directly related to its security posture. Here there was a need for a joint European reply. It was clearly proven that, from a competitive point of view, Europe should engage in these activities if it wished to keep abreast of space developments.' [103]

In essence then, the colloquy drew together a number of interested parties who collectively endorsed the notion that Europe should move towards the establishment of a European military space policy and the detail included therein would from the basis for the WEU Assembly's discussion of the topic in the months ahead.

Over the next year or so, consideration within the WEU Assembly, and a number of its Committees, was to focus in particular on the Strategic Defence Initiative and its implications. The Presidential Committee (which largely directed the agenda of the Assembly) decided that the Assembly's December 1985 meeting should include a wide-ranging debate on the Strategic Defence Initiative. Accordingly, at that meeting, three separate reports on various aspects of the impact of SDI were presented: 'WEU and the Strategic defence initiative - the European pillar of the Atlantic Alliance', submitted by the General Affairs Committee;[104] 'WEU and the Strategic defence initiative (Defence aspects)', submitted by the Committee on Defence Questions and Armaments;[105] and 'WEU and the Strategic defence initiative -guidelines drawn from the colloquy on the space challenge for Europe (proposals)', from the Committee on Scientific, Technological and Aerospace Questions.[106] Collectively, these documents did indeed form the basis of wide-ranging consideration of SDI, examining the major aspects and implications of this programme, with particular reference to the European dimension.

The report of the Committee on Defence Questions and Armaments was principally concerned with defence aspects of the programme. It was based, in large part, on an extensive series of interviews which had been conducted by the Rapporteur throughout January and February of that year, in Bonn, The Hague, London, Brussels, Paris and the USA, principally with defence ministries, but also with representatives of industries involved, or potentially involved in the programme. In addition, the committee, in July, met with a number of representatives to the Conference on Disarmament in Geneva. The outcome of these meetings was a reasonably comprehensive, if concise, report on a number of aspects of the SDI programme. It covered the history of Ballistic Missile Defence (BMD), from the early steps in the 1960s through to President Reagan's initiative, the thorny question of SDI and arms control before moving on to the European dimension, with an examination of European opinion, and the US offer of European participation.

The Committee's recommendations to Council were, on the whole, positive. It was recommended that the Council should agree a common response to SDI which should: (a) stress the importance of avoiding an arms race in space; (b) accept research compatible with existing arms control agreements and of a nature and scale which will enhance stability and security; and (c) permit European industry to participate on mutually advantageous terms in appropriate areas of SDI research.[107] However, as (b) above suggests, there was an element of concern that the development of such a system should not interfere with arms control. This was strengthened by two of the other recommendations, namely that Council should 'urge the Soviet Union and the United States to ensure that space defensive measures of themselves do not become an obstacle to balanced and verifiable agreements limiting strategic and intermediate range nuclear weapons' and should 'instruct the new agency for the study of arms control and disarmament questions to report annually on the arms control impact of the SDI'. Thus, it is clear that from the outset the WEU assembly shared the concern, which was to become much more widespread over the next few years, that while research and testing of such a BMD capability could be cautiously welcomed, it must not be allowed to proceed at the expense of existing or potential arms control agreements.

The Committee on Scientific, Technological and Aerospace Questions was more directly concerned with Europe's activities in space. It was scheduled to report back on the colloquy on the space challenge for Europe which it had convened in March 1985 in Munich, and had expanded the section on SDI to meet the WEU Assembly President's desire for debate on this topic. In organising the colloquy, the Committee had been mindful of the fact that the previous year, the Council had:

> '... acknowledged that the European space industry is a reality, that its achievements are considerable and that its potential is significant. Furthermore, Europe's international influence, and to some extent its security, will in the long term depend on the position it will occupy in space activities.' [108]

The report reiterated that which the committee had stressed time and again, namely that Europe must preserve an independent European capability in space affairs, and went on to outline the aerospace capabilities of each of the member states and various international activities, including those planned through ESA and examples of the work being done in Europe on the US-led space station programme. Moving on to the specific question of the strategic defence initiative, and European cooperation, the report detailed the political positions which had been outlined by the respective ministers of member states at the colloquy.

The French Minister of Research and Technology, M. Curien was of the opinion that space activities in Europe were necessary in a defence context, but that France considered that such a programme [as SDI] could only lead to the revival of an arms race and that it was therefore undesirable.[109] This view was not shared by Mr Timmermann, Secretary of State for Defence of the Federal Republic of Germany. He disagreed with the French Minister on European participation in SDI and was of the opinion that it was of vital interest for the Federal Republic to take part in the US research effort, and expressed the view that a government to government accord should be signed by the end of the year, or early 1986. His attitude was summarised by the following statement:

> 'Europe cannot stand aside if the world powers intensify their systematic efforts for a military utilisation of space. Therefore we should jointly pursue considerations on a European anti-missile system which, either independently or as an element of SDI would open up space as a dimension of defence. This is all the more true if, as

generally expected, the use of space for defence purposes is going to improve the chances of reducing the danger of war.' [110]

This was a position with which the British essentially agreed. Mr. Pattie, the UK Minister of State for Industry and Information Technology, stated that the UK government had accepted the invitation to participate in the research phase of the programme. He further pointed out that

> 'A programme with the size and complexity that SDI appears to have will bring a great deal of civil spin-off from the research which will have to be done. The United Kingdom will be looking for a genuine cooperative role and a sharing of the necessary technology.' [111]

The Italians, through their Minister for Research and Technology, Mr Granelli, stressed the need for a joint position on participation in SDI to be adopted by the member countries of WEU. Their position was outlined as onc of general ambivalence, pointing out that Italian participation in the implementation of the SDI programme should not be interpreted as anticipating the general political and strategic efforts which the European countries in particular will have to assess carefully.

> 'It is therefore unacceptable to exclude as a matter of principle participation in a common research stage for space defence. But at the same time we must reject any claim to move automatically onto the implementation of a system when we do not even know how it will really work. It is moreover well known that Italy, in this connection, considers the limits laid down in the ABM treaty cannot be transgressed.' [112]

In addition to these views expressed by the politicians, the report also reflected the views of other groups, including industrialists and scientists. The final recommendations, reiterating in other areas of space broad suggestions similar in tone and content to many which had been made before, such as the consideration of the adoption of a coherent space programme, and to help the

European aerospace industry reorganise to ensure independence and competitiveness in the international market, produced three recommendations to Council, which had a direct bearing on SDI.

They recommended that Council should urge member governments to:

- define the cooperative framework in which the defence aspects of European space activities can be discussed and determined;
- accept non-aggressive applications of military space technology such as communications, surveillance, navigation and the use of satellites for crisis management and treaty verification to strengthen strategic stability in relations between NATO and Warsaw Pact countries as indicated in the NATO statement of 8th January 1985;
- to pursue jointly research on a European anti-missile system independently or as part of SDI [113]

While the broad thrust of these recommendations is not surprising, there is one very significant point. The second of the three recommendations cited above contains reference to the use of satellites for crisis management and treaty verification. As issue areas which have grown in importance since the demise of the Cold War, these functions in particular were to be firmly grasped by WEU in subsequent years and formed part of the fundamental rationale for the ultimate establishment of the WEU satellite centre.

The Report of the General Affairs Committee was, as its title ('WEU and the strategic defence initiative - The European pillar of the Atlantic Alliance') suggests, concerned far more with the role of the WEU itself. Indeed, as the introduction to the report explains, the report was originally on the topic of the European pillar of the Alliance, and was modified to become part of the desired debate on SDI. [114] In some ways, therefore, while containing a section on SDI, and thus legitimately featuring in the Assembly's consideration of this topic, it is, in itself, largely focused on the much broader topic of the future role of the WEU as an organisation. To this extent, it is a counterpoint to the detailed consideration of SDI, or even space in general, giving a contemporary assessment of the potential role of the WEU in the realm of European and transatlantic security.

It opens with an interesting comment on the very term 'European pillar', before going on to define the reasons for the reactivation of the WEU as a body and how Europe's participation in the defence of Europe might be organised. For many years, according to this report, the governments of the European member countries of the Atlantic Alliance have avoided using the term 'European pillar' of the alliance, lest this should indicate a widening gap between the two shores of the Atlantic. As the report points out, the very idea of an Atlantic community based on two pillars is one which originated on the American side of the Atlantic, being referred to for the first time in a speech by President Kennedy in Philadelphia in April 1962.[115]

The report goes on to suggest that there are several reasons for the reactivation of WEU and the increased European participation in defence of the continent which was emerging. Some of these were as contained in the Council's Rome declaration of 27 October 1984:

- a better utilisation of WEU would not only contribute to the security of Western Europe but also to an improvement in the common defence of all the countries of the Atlantic Alliance and to greater solidarity among its members
- there are specifically Western European geographical, political, psychological and military dimensions
- increased cooperation within WEU will also contribute to the maintenance of adequate military strength and political solidarity and, on that basis, to the pursuit of a more stable relationship between the countries of East and West by fostering dialogue and cooperation
- there is a need to make the best use of existing resources through increased cooperation, and through WEU to provide a political impetus to institutions of cooperation in the field of armaments [116]

It is clear from this, that the WEU was intending to provide a Western European focus for security and defence considerations. Through the development of the concept of burden-sharing attempts had been made to ensure that both sides of the Atlantic Alliance contributed to the general defence of Europe, but there had always been a fear amongst European partners that the US might, at some indeterminate future stage, seek to uncouple its defence interests from those of the European allies. In order to counter this fear, a number of US guarantees had been obtained over the years, including the permanent stationing of strong US forces on mainland Europe and the development of consultations within the NATO framework between the allies on political and military matters. Nevertheless, as this report indicates, there was a strong feeling that the situation would be further improved if the European partners could agree among themselves, and consult with each other, in order to gain greater weight within the Alliance as a whole. This was in part dealt with by the Rome declaration which served as the basis for the reactivation of the WEU. Rome had agreed that Ministers of Defence of the WEU member states would attend Council meetings and that matters relating to Alliance strategy would be one of the focuses of the WEU's work.[117] Thus it was hoped to make it clear that the role of the WEU was firmly envisaged as being complementary to that of the NATO alliance, not in competition with it.

The elements of the report more specifically related to the SDI programme, suggested that the current trends in US defence policy, and the announcement of the SDI programme in March 1983, provided a new impetus and added urgency for organising the European pillar of the Alliance. The report is quite clear about the political implications of this move:

> 'The fact that the SDI stems from a purely American decision, in which the United States' allies were asked to take no part, can but remind the European members of the Atlantic Alliance of the need for better consultations and for making their voice better heard to defend their own security interests. It makes it even more urgent to develop the various elements ... for allowing a European pillar to be formed in the alliance.' [118]

The report goes on to stress how disastrous it would be if a 'deep rift' were to form between the European members of the Alliance over the answer to be given to the American proposal, because once again it would leave the United States alone to decide what its allies' strategy should be.

Overall, the combination of these three reports on different aspects of the SDI programme and its impact on Europe, give an interesting perspective on the WEU and the attitude of its members at this time. The overall impression emanating from the three documents is one which portrays the European members of NATO as, while certainly not seeking to dissociate themselves from their transatlantic partners, most definitely trying to find a coherent, collective identity. This was by no means designed to undermine the Atlantic Alliance, but was instead seen as an opportunity for the European members of NATO to enhance their own ability to input into the debate on the conduct of European security. In a sense, it was not even the SDI programme that was at the centre of this entire debate, rather it simply happened to be an issue which came along at the right time and provided the raison d'être the West Europeans had been looking for in their efforts to mould a more collective identity. The whole debate revolved much more around thc role of WEU in the overall European institutional architecture and the redefining of the relationship between the Alliance allies, than it did on the merits or demerits of the SDI programme. As hindsight of course reveals, SDI itself was a programme which was to be constantly redefined, and which would eventually bu dramatically downgraded, while European involvement in it, or indeed the involvement of any other state, was to say the least, very limited.

Thus, in one sense, the deliberations of the WEU Assembly in 1985 are irrelevant, in that they did not result in any serious European involvement in the SDI programme. However, they are of considerable significance in that consideration of this issue provided a focus for the more general consideration of the WEU and its role in the international arena. Once again, it was largely a space-related topic which was bringing this issue to the attention of the member states. As has been mentioned above, it raised the question of tasks such as crisis monitoring, and treaty verification which were to remain on the WEU agenda for the next decade, and would ultimately, at least in part, lead to the establishment of the WEU satellite centre. In this regard, therefore, the deliberations and views expressed in this particular series of reports, did indeed have long-term implications for the organisation as a whole, and the general framework of Western European security.

There were, however, to be a number of other reports which continued the emphasis which the WEU, and the WEU alone at that time, placed on the role of European space activity.

April 1986 saw a report from the Committee on Scientific, Technological and Aerospace questions with the almost eponymous title, 'Scientific, Technological and Aerospace questions and Western European defence'.[119] This examined the defence related uses of several advanced technologies, looking in broad terms at areas such as information technology, computers and semiconductors. It also, however, touched on space activity and describing 1986 as one of the most crucial years for European space activities, following on from the ESA ministerial meeting of the previous year, as well as commenting on Europe's important role in space, something which it outlined as demanding a strong European effort as well as effective cooperation with the United States.[120] Of much greater significance, however, was the report issued the following year, in April 1987, which represented a major assault on the space front.

Revealingly titled, 'European space policy until 2000',[121] this report was a wide-ranging and detailed assessment of existing and proposed space activities. It opened with an exposition of the rationale for civilian space efforts and the distinction between civilian and military programmes, before going on to examine in some detail developments outside Europe, including summaries of activities in the international commercial launch market and in international cooperation, before finally outlining European space programmes up to the turn of the century. The most significant recommendations of this document were that European space cooperation, which had already made Europe the third space power in the world, should be strengthened, and, as a pointer to the clear desire which existed to move this programme forward, it also called for an examination of the repercussions associated with the establishment of a European military programme for communications, navigation, observation and reconnaissance satellites.[122]

This recommendation, can be seen, in conjunction with those which had gone before, as the genesis of the programme which was ultimately to result in the WEU embracing an operational military space role. The constant reference to the importance of military space activity and policy which had emanated over the years from the WEU Assembly, resulted, in 1989, in the Council mandating an ad hoc group on space to consider such issues, in particular the potential use of space in relation to the implementation of arms control agreements. The developments which were to ensue from this historic decision, are considered in the following chapter, on the creation of the WEU Satellite Centre and its role.

ENDNOTES

[35] Assembly of WEU, Eighth Ordinary Session (First Part), Report submitted on behalf of the Committee on Defence Questions and Armaments, *European Space Organisation*, Document 232, (Assembly of WEU, Paris, 3 May 1962)

[36] Assembly of WEU, Document 232

[37] Assembly of WEU, Seventh Ordinary Session (Second Part), Report submitted on behalf of the Committee on Defence Questions and Armaments, *Report on the state of European Security*, Document 215, (Assembly of WEU, Paris, 10 November 1961)

[38] Assembly of WEU, Document 232, p.160

[39] Assembly of WEU, Recommendation 69, cited in Tenth Ordinary Session (Second Part), Report submitted on behalf of the Committee on Defence Questions and Armaments, *Progress in space technology and possible repercussions on strategy*, Document 322, (Assembly of WEU, Paris, 9 November 1964) p.3

[40] Assembly of WEU, Document 322

[41] Assembly of WEU, Document 322, p.9

[42] Assembly of WEU, Document 322, p.2

[43] Assembly of WEU, Eleventh Ordinary Session (Second Part), Report submitted on behalf of the Committee on Space Questions, *State of European space activities: the political choice*, Document 358, (Assembly of WEU, Paris, 27 October 1965)

[44] Assembly of WEU, Document 358, p.2

[45] Assembly of WEU, Document 358, p.7

[46] Assembly of WEU, Document 358, p.8

[47] Assembly of WEU, Document 358, p.8

[48] Assembly of WEU, Document 358, p.2

[49] Assembly of WEU, 12th Ordinary Session (First Part), Report submitted on behalf of the Committee on Space Questions, *State of European space activities. The ELDO crisis*, Document 371, (Assembly of WEU, Paris, 23 May, 1966)

[50] Assembly of WEU, Document 232

[51] Assembly of WEU, Document 371, p.3

[52] Assembly of WEU, Document 371, p.4

[53] Assembly of WEU, Document 371, p.4

[54] Assembly of WEU, Document 371, p.2

[55] Assembly of WEU, 12th Ordinary Session (First Part), Report submitted on behalf of the Committee on Space Questions, *Global communications by satellite - Europe's role*, Document 372, (Assembly of WEU, Paris, 23 May, 1966)

[56] Assembly of WEU, Document 372, p2

[57] Assembly of WEU, 12th Ordinary Session (Second Part), Report submitted on behalf of the Committee on Space Questions, *Juridical problems and space policy*, Document 388, (Assembly of WEU, Paris, 21 November, 1966)

[58] Assembly of WEU, 12th Ordinary Session (Second Part), Report submitted on behalf of the Committee on Space Questions, *State of European space activities*, Document 389, (Assembly of WEU, Paris, 21 November, 1966)

[59] Assembly of WEU, Document 389, p.3

[60] Assembly of WEU, Document 389, p.5

[61] Assembly of WEU, Document 389, p.4

[62] Assembly of WEU, 13th Ordinary Session (First Part), Report submitted on behalf of the Committee on Scientific, Technological and Aerospace Questions, *State of European space activities. European space and satellite technology and international competition*, Document 402, (Assembly of WEU, Paris, 2 May, 1967)

[63] Assembly of WEU, 13th Ordinary Session (Second Part), Report submitted on behalf of the Committee on Scientific, Technological and Aerospace Questions, *State of European space activities*, Document 429, (Assembly of WEU, Paris, 17 November, 1967)

[64] Assembly of WEU, Document 429, p.2

[65] Assembly of WEU, 14th Ordinary Session (First Part), Report submitted on behalf of the Committee on Scientific, Technological and Aerospace Questions, *The Soviet orbital bombardment system*, Document 444, (Assembly of WEU, Paris, 3 May, 1968)

[66] Assembly of WEU, Document 444, p.7

[67] Assembly of WEU, Document 444, p.8

[68] Assembly of WEU, 15th Ordinary Session (Second Part), Report submitted on behalf of the Committee on Scientific, Technological and Aerospace Questions, *State of European space activities - INTELSAT*, Document 495, (Assembly of WEU, Paris, 25 November, 1969)

[69] Assembly of WEU, Document 495, p.2

[70] Assembly of WEU, 17th Ordinary Session (Second Part), Report submitted on behalf of the Committee on Scientific, Technological and Aerospace Questions, *Prospects of space*

collaboration between the United States and Europe, Document 562, (Assembly of WEU, Paris, 29 November, 1971)

71 Assembly of WEU, Document 562, p.8

72 Assembly of WEU, Document 562, p.9

73 Assembly of WEU, Document 562, p.2

74 McLean, A., *Western European Military Space Policy* (Dartmouth Publishing, Aldershot, 1992), p.69

75 Assembly of WEU, 18th Ordinary Session (First Part), Report submitted on behalf of the Committee on Scientific, Technological and Aerospace Questions, *Orbital stations and the civil and military consequences, Document* 572, (Assembly of WEU, Paris, 8 May, 1972)

76 Assembly of WEU, Document 572, p.3

77 Assembly of WEU, Document 572, p.10

78 Assembly of WEU, Document 572, p.2

79 Assembly of WEU, 20th Ordinary Session (First Part), Report submitted on behalf of the Committee on Scientific, Technological and Aerospace Questions, *The European Space Agency. Reply to the Nineteenth Annual Report of the Council*, Document 639, (Assembly of WEU, Paris, 21 May 1974)

80 Assembly of WEU, Document 639, p.3

81 McLean, A., *Western European Military Space Policy* (Dartmouth Publishing, Aldershot, 1992), p.72

82 Assembly of WEU, 21st Ordinary Session (Second Part), Report submitted on behalf of the Committee on Scientific, Technological and Aerospace Questions, *United States - European cooperation in advanced technology*, Document 687, (Assembly of WEU, Paris, 17 November, 1975)

83 Assembly of WEU, 24th Ordinary Session (First Part), Report submitted on behalf of the Committee on Scientific, Technological and Aerospace Questions, *United States-European cooperation and competition in advanced technology*, Document 773, (Assembly of WEU, Paris, 17 May, 1978)

84 Assembly of WEU, Document 773, p.3

85 Assembly of WEU, 24th Ordinary Session (First Part), Report submitted on behalf of the Committee on Scientific, Technological and Aerospace Questions, *Application satellites*, Document 766, (Assembly of WEU, Paris, 2 March, 1978), and also Assembly of WEU, 24th Ordinary Session (Second Part), Report submitted on behalf of the Committee on Scientific, Technological and Aerospace Questions, *Application satellites. Part II*, Document 784, (Assembly of WEU, Paris, 26 October, 1978)

86 Assembly of WEU, Document 766, p.2

87 Assembly of WEU, Document 784

88 Assembly of WEU, 26th Ordinary Session (First Part), Report submitted on behalf of the Committee on Scientific, Technological and Aerospace Questions, *A European earth resources detection satellite programme*, Document 842, (Assembly of WEU, Paris, 29 April, 1980)

89 Assembly of WEU, Document 842, p.2

[90] Assembly of WEU, 27th Ordinary Session (First Part), Report submitted on behalf of the Committee on Scientific, Technological and Aerospace Questions, *Future of European space activities - Reply to the Twenty-Sixth Annual Report of the Council*, Document 883, (Assembly of WEU, Paris, 15 June, 1981)

[91] Assembly of WEU, Document 883, p.7

[92] Assembly of WEU, Document 883, p.2

[93] Assembly of WEU, 30th Ordinary Session (First Part), Report submitted on behalf of the Committee on Scientific, Technological and Aerospace Questions, *The military use of space*, Document 976, (Assembly of WEU, Paris, 15 May, 1984)

[94] Assembly of WEU, Document 976, p.4

[95] Assembly of WEU, Document 976, p.4

[96] Assembly of WEU, Document 976, p.5

[97] Assembly of WEU, 27th Ordinary Session (First Part), Report submitted on behalf of the Committee on Scientific, Technological and Aerospace Questions, *Future of European space activities - reply to the twenty-sixth annual report of the Council*, Document 883, (Assembly of WEU, Paris, 15 June, 1981)

[98] Assembly of WEU, 30th Ordinary Session (First Part), Report submitted on behalf of the Committee on Scientific, Technological and Aerospace Questions, *The military use of space*, Document 976, (Assembly of WEU, Paris, 15 May, 1984) p.7

[99] Speech by President Mitterand in the Second Chamber of the States-General, The Hague, 7 February 1984, cited in Assembly of WEU Document 976, p.16. (Appendix 1)

[100] Assembly of WEU, 30th Ordinary Session (Second Part), Report submitted on behalf of the Committee on Scientific, Technological and Aerospace Questions, *Military use of space Part II*, Document 993, (Assembly of WEU, Paris, 8 Novenber, 1984)

[101] Assembly of WEU, Document 993, p.3

[102]Gibson, R., *Official Record of Colloquy on the Space Challenge for Europe*, 18-20 September 1985, (Office of Clerk of the Assembly of WEU, Paris, 1985), p.99

[103] Lenzer, C., *Official Record of Colloquy on the Space Challenge for Europe*, 18-20 September 1985, (Office of Clerk of the Assembly of WEU, Paris, 1985), p.102

[104] Assembly of WEU, 31st Ordinary Session (Second Part), Report submitted on behalf of the General Affairs Committee, *WEU and the Strategic defence initiative - the European pillar of the Atlantic Alliance*, Document 1034, (Assembly of WEU, Paris, 5 November, 1985)

[105] Assembly of WEU, 31st Ordinary Session (Second Part), Report submitted on behalf of the Committee on Defence Questions and Armaments, *WEU and the Strategic defence initiative (Defence Aspects)*, Document 1033, (Assembly of WEU, Paris, 4 November, 1985)

[106]Assembly of WEU, 31st Ordinary Session (Second Part), Report submitted on behalf of the Committee on Scientific, Technological and Aerospace Questions, *WEU and the Strategic defence initiative - guidelines drawn from the colloquy on the space challenge for Europe (Proposals)*, Document 1036, (Assembly of WEU, Paris, 6 November, 1985)

[107] Assembly of WEU, Document 1033, p.6

[108]Assembly of WEU, Document 1036, p.3

[109] Assembly of WEU, Document 1036, pp10-11

[110] Assembly of WEU, Document 1036, pp11-12

[111]Assembly of WEU, Document 1036, p. 12

[112] Assembly of WEU, Document 1036, p. 12

[113] Assembly of WEU, Document 1036, p. 2
[114] Assembly of WEU, Document 1034, p.3
[115] Assembly of WEU, Document 1034, p.3
[116] Assembly of WEU, Document 1034, p.3
[117] Assembly of WEU, Document 1034, p. 6
[118] Assembly of WEU, Document 1034, p.10
[119] Assembly of WEU, 32nd Ordinary Session (First Part), Report submitted on behalf of the Committee on Scientific, Technological and Aerospace Questions, *Scientific, Technological and Aerospace questions and Western European defence*, Document 1055, (Assembly of WEU, Paris, 28 April, 1986)
[120] Assembly of WEU, Document 1055, p.2
[121] Assembly of WEU, 33rd Ordinary Session (First Part), Report submitted on behalf of the Committee on Scientific, Technological and Aerospace Questions, *European Space Policy until 2000*, Document 1098, (Assembly of WEU, Paris, 29 April, 1987)
[122] Assembly of WEU, Document 1098, p.4

CHAPTER FOUR

THE ESTABLISHMENT OF THE WEU SATELLITE CENTRE

I. From Rhetoric to Reality

As has already been outlined in some detail in the preceding chapter, the Western European Union, and in particular its parliamentary assembly, have in fact had a long-standing interest in space activities, dating back to their first report on European space organisation in May 1962.[123] Such interest, unique among European international organisations, in both the military and civilian aspects of space activity, has meant that the foreign and defence ministers of the member countries who comprise the WEU Council at Ministerial level have periodically been reminded of the value and importance of space. Indeed, the involvement and interest of the WEU in the uses of space represents a permanent thread running through the history of the organisation, generally aiming to consolidate and intensify space cooperation within Europe.

The Ministers first seriously discussed the prospect of space cooperation within the context of the WEU in April 1989 when they tasked an ad hoc group to study the possible uses of space technology in three specific areas - the verification of arms control, the monitoring of crises affecting European security, and the monitoring of environmental hazards. The report of the ad hoc study group was duly presented to the WEU ministers in November 1989, suggesting that, 'in the security field, European space cooperation had not reached a level commensurate with its potential.'[124] There followed, at the request of the Ministers, a further analysis of various aspects of such cooperation, which was paralleled by a series of WEU Assembly reports and recommendations on the possible development of a European space-based observation system.[125] The following year, in December 1990, the WEU Council, considered that a step by step approach to setting up such a system was the appropriate route to take,[126] and, meeting in Vianden, Luxembourg on 27 June 1991 agreed to set up a satellite data interpretation centre for an experimental period of three years.[127] The immediate task of this centre was 'to train European experts in the photo-interpretation of satellite-derived data, to compile and process accessible data and to make those data available to member States, particularly within the framework of the verification of arms control agreements, crisis monitoring and environmental monitoring.'[128]

The Council, at the same meeting, also tasked the ad hoc Sub-Group on Space to pursue studies on the possibilities for medium and long-term cooperation on a European satellite observation system. Created under the auspices of the WEU Council, the Ad Hoc Study Group on Space is a small study group which reviews activities in the field of space technology, and, more specifically, defines prime areas for co-ordinated activity. Furthermore, the linkage between space cooperation and arms control verification, including verification of the CFE Treaty was underlined at the Vianden meeting.[129] The combination of these decisions was commonly regarded as representing a significant step towards the emergence of the WEU as an active body, with practical, operational capabilities in the field of space-based reconnaissance and surveillance.

The 1991 Vianden ministerial decision can therefore be seen as a turning point in the overall development of a coherent, positive strategy in relation to Western European military space policy. This agreement to establish a multinational centre under the auspices of the WEU was of importance for three reasons. Firstly, the major states of Western Europe had, for the first time, taken concrete steps to collaborate and cooperate in the use of space to enhance their security - which in itself was a significant step in the overall utilisation of space for security. Secondly, such a move can also be regarded as a major step towards the overall Europeanisation of defence and security. Rather than continuing to rely almost solely on the USA for space derived intelligence data, the European states were indicating their recognition of the fact that they might require their own capabilities in such an area.

Finally, and perhaps most importantly in the longer term, it represented a commitment to a significant enhancement of the WEU as an operational body within the overall field of European security. While the WEU had undertaken a number of operations, and established a Planning Cell in Brussels, the Satellite Centre would be the first established operational body directly under the control of the multinational WEU.

II. The Birth of the Satellite Centre

The Satellite Centre is a WEU subsidiary organ, in accordance with Article VIII, paragraph two, of the 1954 modified Brussels Treaty, which permits the Council of WEU to set up such subsidiary bodies as may be considered necessary. In defining the status of the Centre in this manner the Satellite Centre gained a certain amount of independence, within the overarching framework of the WEU. Unusually, the Centre was to have its own, separate budget, which, for the original three year experimental phase was set at 38.25 million ECUs. This comprised a general budget of 30.6 million ECUs and an additional 25 per cent to allow for contingencies. Moreover, budgetary costs were to be shared proportionately amongst the WEU members on the basis of a pre-agreed cost-sharing formula with France, Germany, Italy and UK each contributing 16.75 per cent. The overall budget, and, more generally, the activities of the Centre, were scheduled to be reviewed in 1994 or 1995, at which time member states would have the option of reconsidering their participation in the project. The annual budget of the Centre, allocated each year by the Council, covers such things as staff salaries, travel, operation costs and maintenance of the premises, as well as the capital and running costs of the computer system, which is at the heart of the Centre's operations. It also includes provision for the purchase of space images and the maps the Centre needs for its work.

The overall management of the Centre is the responsibility of its director and a steering committee on which each member state has one representative. The role of the steering committee was to lay down technical and financial guidelines to be implemented by the Centre, and to be in constant liaison with the national authorities calling upon the Centre's services.[130]

In November 1991 a site for the new satellite centre was eventually selected at Torrejon in Spain. The location of the facility was finalised at the Bonn meeting of the Ministerial Council. During the year there had been considerable debate as to the selection of a suitable site with Farnborough in the UK, Toulouse in France, and Torrejon in Spain all under consideration. The ultimate decision to select Torrejon was to the considerable delight of the Spanish, who regarded such a move as concrete confirmation of their acceptance into the WEU, a body which they had only joined in 1988 (with their entry only being ratified in 1990).

The 3400m^2 site at Torrejon de Ardoz - a former United States Air Force base - was favoured primarily due to the fact that many of the existing buildings could quickly be renovated to house the Satellite Centre. Similarly, the location was preferred in that not only was it only 20km from Madrid, but was also in close proximity to the headquarters of INTA - the organisation responsible for the Spanish space programmes. The centre officially came into being on 1 January 1992, with its structure and budget being approved in March of that year. In December, the formal agreement was signed between the Spanish government and the WEU concerning the provision of the site and the building, with the annual rent being agreed with Spain at $116,000 for a minimum fixed period of 10 years. Following total refurbishment of the buildings to house the Satellite Centre, it was ready for its new occupants to move in after just five months. The formal inauguration ceremony was duly held in April 1993, when the Centre was officially declared open by Julian Garcia Vargas, the Spanish Defence Minister, in the presence of, amongst others, Willem van Eekelen, Secretary-General of the WEU. By that time some 60 per cent of the staff were in place, along with the first segment of the technical equipment. By January 1994, all the equipment and staff were in place.

In terms of its organisational structure, the Satellite Centre consists of a directorate, four divisions and a financial section. The four divisions are the operational division (Exploitation), which comprises nearly half the total staff complement of 59 national representatives from the 10 member states (as at 1 January 1996), and three support divisions (Technical, Research and Administration).[131]

Mission Objectives

The WEU satellite centre had essentially four objectives during its three-year experimental phase:

1. To demonstrate the application of space imagery for the purpose of treaty verification, crisis monitoring and environmental monitoring
2. To train analysts from the member states
3. To develop computer techniques applied to the interpretation of images
4. To supply initial operational imagery interpretation products for verification, crisis monitoring and environmental monitoring

Each of these four objectives forms a necessary part of the overall experiment. For the satellite centre is not just an experiment in technological terms. As the first Director of the Centre (Barry Blaydes, of the UK) described it:

> '... it is very important to understand that the experimental concept involves much more than just the Centre. It involves an experiment with all the users that will use the Centre so that at the end of this period they understand the potential benefits of the Centre and can therefore advise their ministers accordingly. We must achieve

feedback, through the users, to the nations to support the next round of decision making.' [132]

This raises the question of exactly who the users of the WEU Satellite Centre were actually intended to be. In fact, when it was established in its experimental phase, it was not envisaged that the sole user of the facility would be the WEU itself. It was always envisaged that it would be open to tasking by other users as well. In fact, three distinct categories of users were identified at the outset. The primary user, and that which takes priority over all others, was of course always intended to be the Council of WEU and other WEU organisations, such as the Planning Cell located in Brussels. The second category of user was member states of the WEU who wished to make use of the Satellite Centre capabilities to help with their decision-making as members of the WEU. The third category was member states of the WEU in order to satisfy their own, national requirements. In fact the Centre may be tasked by any of the ten full member states of WEU and also by the three Associate members: Iceland, Norway and Turkey.

These three categories, in priority order, formed the envisioned user base of the Centre. Additionally, however, it was agreed that it would be possible for bodies external to the WEU to task the Satellite Centre if they wished to do so, always subject to the consent of the WEU Council, who would consider the appropriateness of the task. Falling into this category would be such organisations as the Organisation for Security and Cooperation in Europe (OSCE) or the United Nations, for example. There was one important *caveat* attached to this - and indeed to the use of the Centre by individual WEU member states for national purposes - the work carried out by the Centre was open to all WEU member states. In other words, all products and reports which were produced by the Centre would be delivered to the particular user who had requested them, but they would also be available to any WEU body and all the WEU member states. Thus, as Barry Blaydes described it, 'there is no question of the Centre undertaking particular, specialised work which any one nation only wishes the results to be kept to itself. Those are the proposed conditions for the operation of the Centre'.[133] This rider fully establishes the multinational nature of the venture. It effectively precludes use of the Centre for solely national interests which any member state did not wish its fellow members to be aware of.

More general political control of the Centre stems from two sources, or mechanisms:

> 'The Centre only receives tasks from a nation via a single point of contact, known as a national cell, who has been formally nominated by the country concerned. National political control is maintained by this means. In addition, all tasks, whether originating from WEU itself or from member states, are considered by the Permanent Council.' [134]

In 1995 it was revealed that an updated mission statement for the Centre was being proposed in the light of the further elaboration of the role of the WEU itself over the preceding three years.

Director of the Centre, Barry Blades said:

> 'I expect the final Mission Statement will include aspects such as humanitarian and rescue tasks, crisis management (which is very different from crisis monitoring), peacekeeping tasks and also maritime surveillance ... the Mission Statement will be adjusted in the context of treaty verification and is likely to read: to provide support to treaty verification and support to arms control and proliferation.' [135]

Unfortunately the new Mission Statement is classified and as a result cannot be reproduced.[136] However, we can assume that these tasks are at the heart of the activities envisaged for the Centre as it develops its role over the next few years.

Technical Capabilities

In essence, the WEU satellite centre requires the technical capacity to acquire, store, retrieve and analyse data from a range of sources in order to achieve its mission. Procurement of the necessary technical equipment was undertaken through a multinational project - the Falcon Consortium - headed by the UK-based Cray Systems.

In terms of input, the Satellite Centre as it is presently established remains a ground-based organisation, with no space-based assets of its own. Consequently, the data which it interprets must originate from some other source. It currently utilises data commercially obtained from the US Landsat, the French Spot series of satellites, and the European Space Agency's ERS-1 remote sensing satellite, which has all weather capabilities. Consideration has also been given to the possibility of utilising Russian and Ukrainian imagery. In addition, a Memorandum of Understanding was signed in 1993, the day before the satellite centre was officially inaugurated, to give the Centre access to imagery obtained by the joint French/Italian/Spanish Helios series of military observation satellites, the first of which, Helios 1A, was launched in July 1995. The first data from this satellite was available to the Satellite Centre from 7 May 1996.

The whole question of the supply of imagery was addressed, and neatly summarised, by the then director of the Centre, Barry Blaydes, addressing the WEU Colloquy on a European space-based observation system in 1995. He pointed out that as director of an image interpretation centre securing a comprehensive source of supply of imagery was his prime concern. He suggested that there were four vital characteristics needed of the overall source of imagery - that it be flexible, reliable, effective and efficient. By flexible, he was referring to the fact that the supply must offer an acceptable compromise to the conflicting requirements of geographic coverage, response times, data cut off date and image quality. In order to achieve such flexibility, a range of different sources, allied to a comprehensive and readily accessible archive on-site, of previously obtained imagery offers the best solution. In terms of reliability, he was referring to the political aspect. There had to be a high probability that the source of imagery would not be cut off at times when the interests of the owner of the source do not coincide with the needs of the centre - a strong argument for having one's own, independent source of imagery. Effectiveness refers to the actual imagery being suitable for the purposes of the Satellite Centre - that it be in a format which can be readily used, for example. And efficiency refers to the ease with which the exchange between the supplier and the Centre can be arranged.[137]

Such considerations clearly indicate the relative complexity of the range of factors which constantly have to be taken into account in the timely and appropriate acquisition of satellite imagery for analysis. They also do not take into account one further very important factor, that of the cost of such acquisitions, a point which will be expanded upon later in this chapter.

Satellite data is, of course, only one data source which the centre can utilise. It also has the capability to obtain and make use of other forms of data, e.g. airborne imagery, maps etc. Such data is stored using an electronic database, of commercial origin, with optical disk being the main storage medium. When data is acquired it is inspected for image quality and then entered into the on-line library, which contains all the other elements of the Centre's data base, including digitised maps, treaty data and the results of previous interpretation work in the form of diagrams and reports. The database is accessed by way of one of a number of workstations which can also output the data to a variety of peripheral devices, such as printers.

The overall system architecture of the Satellite Centre utilises two separate, but complementary software systems for imagery exploitation: ERDAS for general scientific applications and OCAPI, used more specifically for defence purposes.[138]

In order to carry out the actual work of interpretation of the data the images are transferred to one of eight double screen, high definition image processing work stations. Using these workstations, the interpreters make use of a variety of software tools to manipulate and enhance the images, to undertake measurements and to create appropriate output, such as annotated diagrams of the site being analysed. There are a further nine workstations for quality control, network management and database consultation, and a further two workstations specifically dedicated to the exploitation of the high definition imagery obtained from the Helios satellite system[139]. All of the system facilities are interconnected by a high speed network, which also allows data to be transferred to the photo-lab wherein the digital data is transferred to film using high resolution laser film writers. These films are then processed and the end product images produced within this lab.

Over time, a considerable quantity of data will be accumulated and stored for future use. If necessary, in response to a particular tasking request, specific data will be purchased. But such data, if it is to be useful, must be categorised in such a manner that it can be readily retrieved. This has been achieved by the logical division of the data into 15 different data classes, which permits data selection on a wide range of criteria, including factors such as geographical area or date of acquisition. Utilising this type of sorting system is designed to offer the speediest and most efficient retrieval of historical data as and when required for future analysis.

The primary means of output of the Satellite Centre to a specific task request is in the form of a dossier. Once a task request has been received the analyst responsible will collate all the relevant imagery and data, carry out the function of interpretation and analysis and compile the dossier to be sent to the body originating the task. The dossier contains the space imagery and maps used for the mission and, above all, maps with information and notes provided by the image interpreter, together with his analytical report. As the Centre itself points out: ' ... this human contribution is the true value added by the Centre to the original space image and it is what the Centre really stands for.' [140]

Overall, therefore it can be seen that the experimental phase of the Satellite Centre was designed to fulfil a variety of purposes. Objectives two, three and four, identified above, namely the training of analysts from the member states; the development of computer techniques applied to the interpretation of images; and the supply of operational imagery interpretation products for verification, crisis monitoring and environmental monitoring were all, in a sense, the easiest to achieve. These all share the common denominator of being practical, applied tasks, the achievement of which could be comparatively easily measured. In addition, they all share the characteristic of being comparatively easy to demonstrate as being useful to the member states.

None of the member states was to prove to be averse to the second objective of training a number of analysts from the individual member states.

Such an objective, regardless of the success or failure of the overall experimental phase of the project, would clearly be of benefit to each of the member states concerned. At the end of the day, even if nothing else came of the project, each state would have acquired a number of trained data analysts who could then be of use at the national level. A similar rationale could obviously appertain to the second objective - that of developing computer techniques which could be applied to the interpretation of images. Again, such techniques could, in the event of the Centre as a whole proving unsuccessful, be of use to the individual member states. The fourth objective - the supply of initial operational imagery interpretation products which could be used for verification, crisis monitoring and environmental monitoring was, in a sense, the measure of the success of objectives two and three. This was the outcome which would prove the success of the training of the analysts and the new computer techniques of analysis which the Centre was supposed to develop. Thus, this would produce hard evidence of the viability and utility of the operation as a whole.

Far and away the most political, and the most crucial objective was the first - to demonstrate the application of space imagery for the purpose of treaty verification, crisis monitoring and environmental monitoring. This, above all the others, would be the paramount objective if the Satellite Centre was to have any chance of moving from its initial experimental status to becoming a permanent body of the WEU. In addition to being almost certainly the most important of the four objectives, it was also the most difficult to achieve, being considerably more abstract than the others. While it would be relatively simple to measure the success of the rather more practical objectives two three and four, demonstrating the application of intelligence derived from space imagery is a much more subjective task. Implicit within this objective are a number of subsidiary objectives, which although unstated are absolutely crucial to the success of the whole venture. In the first instance, to demonstrate the application of space imagery intelligence, one must convince the relevant politicians and defence planners - who tend to be rather conservative by nature - of the utility of such expensive intelligence. Assuming success in this aspect of the task, it is furthermore necessary to convince them that within the changed circumstances of international security it is necessary to have the capability to produce such analysis within Europe. Again given the inherent conservatism of the defence community, the existing reliance on US-supplied intelligence had to be overcome. This could be seen as a particular problem in relation to the United Kingdom, given the history of the 'special relationship' between the UK and the USA. Although, perhaps arguably, this relationship has always been rather more special to the British than to the Americans, in this instance it was the British who would require to be convinced that, at least in this aspect of security, the time had come to forge rather closer links with Europe than had previously been the case. Finally, in order to fully satisfy the objective as set out in the establishment of the Satellite Centre the utility of its products had to be demonstrated in the three distinct fields of treaty verification, crisis monitoring, and environmental monitoring.

In attempting to satisfy these conditions, the question of subjectivity would be a distinct problem. The other three objectives could be measured empirically - analysts could be shown to be trained, computer programmes could be demonstrated to show that they achieved the results they were intended to achieve, and operational imagery analysis and products could be delivered to the member states. But, in demonstrating the application of the potential outputs of the Satellite Centre, a case had to be made. The ultimate determinant of the success or otherwise of making such a case could only be measured in the abstract - by waiting to see how many of the users of the Centre and its facilities were convinced by the arguments.

Thus, while on the face of it the Centre Director and his staff were endeavouring to tackle practical problems in ensuring the physical establishment of the Centre, installing the necessary hardware and software and the like, to ultimately produce hard output from the fledgling Centre the paramount task was actually the much more political one of trying to persuade the relevant national authorities of the utility of, and need for, the Centre and the facilities which it would offer both to the WEU as an international organisation and the individual constituent member states.

II. Satellite Centre Operations and Unaddressed Questions

From the initial decision to establish the Satellite Centre, in June 1991, it took almost three years - until March 1994 - for completion of the acquisition of all the equipment necessary for the experimental phase, and completion of the initial training phase of the Centre's work. It must of course be remembered that this was just under one year from the formal inauguration of the Centre's premises. Thus, it can be seen that the establishment process was, in fact, carried out fairly speedily.

As identified earlier in this chapter, the Centre was to be responsible for a variety of different tasks. In 1994, as an example, it was assigned 14 specific tasks. A total of six of these came under the broad heading of Treaty Verification - four in support of CFE and CSCE inspection visits, and two in support of Open Skies missions. Under the heading of crisis monitoring the Centre undertook 13 tasks, five in preparation for humanitarian aid, seven relating to the detection of military movements, and one in support of preventing non-compliance with sanctions. There were considerably fewer tasks in the field of environmental monitoring - two in relation to the control of nature of terrain, and one in control of illicit movements.[141]

While the actual establishment of the centre is generally regarded as a very positive step by the WEU, there have however been a number of criticisms raised, in particular by the Technological and Aerospace Committee of the WEU Assembly. In their report, 'The development of a European space-based observation system - Part II', submitted to the WEU Assembly in November 1993, a number of concerns are raised concerning the equipment and software chosen for use in the centre. The equipment offered by the successful tenderer was intended mainly to train image-interpreters, but offering few operational possibilities. Indeed operations at the Torrejon Centre were delayed by several months as a result of delays in decisions over the initial operating equipment. Additionally, as Assembly Recommendation 555 indicates, the American software chosen is apparently unable to handle Helios imagery, and it is seemingly very difficult for more sophisticated software to be supplied which might allow operational activities to be pursued. The Committee report suggested that, having chosen the equipment and software on the basis of purely economic criteria may well limit the operational capabilities of the centre, rendering it no more than a European school for image-interpretation.[142]

In May 1996 Jacques Baumel, one of the French parliamentarians in the WEU Assembly was Rapporteur for a report on European intelligence policy presented to the Assembly by the Defence committee.[143] This report was highly critical of the Satellite Centre for not fulfilling its initial expectations regarding 'intelligence-related missions'. The report also suggested that the overall image of the Centre left much to be desired, and called upon member states to make better use of the facility. The report was especially critical with reference to the method of receipt of the imagery from its original sources. Requested images which were ordered by the Satellite Centre arrived by regular mail, causing considerable delays in receipt of the imagery, sometimes taking weeks before they were transferred to the ordering party. This has clear implications for the utility of such imagery in, for example, crisis monitoring, where speed is potentially of the essence.

Another significant problem is that of the cost of purchasing imagery - particularly from the only military source, Helios. Christian Lenzer, for example, one of the German parliamentarians involved, has complained that the Satellite Centre only rarely purchases images from this source. Yet, given that the cost of one such image was tentatively agreed at $40,000 it is far from surprising that such images have only been acquired very occasionally.[144] Such a cost permits the Satellite Centre to purchase very few images, perhaps as few as eight per year. It has, in fact, been suggested that pricing difficulties have been exacerbated with France charging even higher amounts for some images. It has been alleged that France has charged the WEU up to $41,142 per image in an attempt to recoup the cost of the venture.[145] Clearly the operational effectiveness of the Satellite Centre must be severely limited if its basic requirement for imagery is significantly constrained by the inevitable economic limitations that such high costs entail.

Clearly, therefore, the initial years of the fledgling Centre were not trouble-free, and did in fact generate a fairly wide range of criticisms. On the one hand, if such criticisms are well-founded, this poses potentially difficult questions for the operational future of the satellite centre which would have to be addressed by the WEU Council if the experiment were to be extended into a fully operational facility, or indeed to be continued beyond the initial three year experimental phase. On the other hand, if they were merely symptomatic of the type of teething problems which might be expected in undertaking the establishment of such a multinational agency, then perhaps they need not be of quite such significance. Perhaps the fairest assessment of the situation to date is that while the initial difficulties have to a large extent been overcome others, such as the overall suitability of software may have longer term implications for the Centre.

In the final analysis however, it seems that the experimental phase of the Centre was successful. Despite such concerns as those indicated above, the future of the Centre was assured when the Ministerial Council met in Lisbon in May 1995. Having examined the progress made in the field of WEU space activities, they approved a decision on the establishment of the Satellite Centre as a permanent WEU body.[146] Thus, in at least one sense the experiment can be deemed a success. The act of granting permanence to the body can in itself be seen as a measure of its success. In announcing this decision, the Lisbon Council also moved the agenda forward. One way of reducing or even eliminating the Satellite Centre's reliance on the expensive purchase of imagery from existing satellite sources, thus perhaps increasing the efficiency and effectiveness of the Centre would be for it to move into the realm of operating its own satellites. The Lisbon Council meeting also approved a decision tasking the Space Group (formerly the Ad Hoc Study Group on Space) to continue its activities, to study proposed approaches to developing WEU's capability to use satellite imagery for security purposes.

IV. A Space Based Element

The other facet of the WEU role in the development of a European military surveillance/reconnaissance satellite capability is the thorny question of the space-based element of such a system. This topic has also been the focus of considerable attention within the WEU over the past few years.

focus of considerable attention within the WEU over the past few years. When the Satellite Centre was first mooted, parallel consideration was given to this aspect, the rationale being that if Europe was to be self-sufficient in satellite intelligence imagery, it was logical that it should have independent access to its own satellite system.

The studies which have considered the possibility of such an autonomous space-based system originated from the Vianden ministerial Council meeting of June 1991 (at which time the decision to establish the satellite centre was taken). The ad hoc Sub-Group on Space was tasked with pursuing 'studies on the possibilities for medium and long-term cooperation on a European satellite observation system'.[147]

The culmination of these studies was a feasibility study in two parts, drawn up by a management team, working alongside the Satellite Centre, comprising a consortium of some thirty European firms, led by the German firm, Dornier. The first part of the study examined the choice of a number of possible configurations, while the second made a detailed technical analysis, including cost estimates, of the configurations best suited to the Centre's needs. These studies were due to be completed in time for the WEU Council to take a decision on whether or not to commence a programme for an independent European satellite system at their meeting in 1995. Accordingly, to facilitate this, in 1994 the Council had tasked the WEU Space Group to prepare a proposal for decision including the preparation of a draft memorandum of understanding containing detailed specifications.[148]

While all such studies, effectively offering advice to ministers are, essential confidential, one report to the WEU Assembly does contain some information as to the requirements identified in the first part of the study relating to the possible configuration of such a satellite system.

According to the conclusions of the first part of the study, the elements of the WEU system would be as follows:

1. The space sector defined as all satellites in orbit belonging to the system, comprising:

- nominal satellites, operating in low orbit (660-570km). This category includes two optical satellites and two synthetic aperture radar satellites;
- small satellites (of the optical and SAR type) operating at very low altitude as complements to the nominal satellites. These small satellites may have a resolution and life-span inferior to that of the nominal satellites (approximately five to seven years), but allow a greater number of images to be taken each day across a predetermined area. The number of these small satellites will depend on their life-span, the tasks assigned to them and their availability;
- data relay satellites in geostationary earth orbit (approximately 36,000km) for transmitting images and other data gathered by the low orbit satellites.

2. The ground sector, including ground-based facilities and installations necessary for control and use of the system: operating processing, archiving and interpretation centres; one or more stations for tracking, telemetry and control of the data relay and other satellites; means of communications capable of connecting all the stations, transmitting data and receiving requests for data from political and military authorities. [149]

This system definition is clearly quite ambitious, but nonetheless pragmatic. Virtually all such elements would require to be present if Europe was serious about the development of an autonomous observation satellite system. Such an enterprise would indeed be a costly one, a fact noted in the Assembly report on the topic which clearly stated in its conclusions that:

> 'The cost of a space-based defence system far exceeds the resources of a single country and a joint space policy is the only means of achieving its implementation'. [150]

Indeed, overall the report is a moderate one. Unlike some of its predecessors, it does not have the same sense of compulsion, rather offering a more guarded call for consideration of the topic as a whole. While the recommendations to Council include commencement of a European space defence policy, taking all aspects of the problem into account; and to take the decision necessary for the development of the satellite centre and for starting work on the main observation system on the basis of a strategic analysis of what is at stake in the medium and long term, it also recommends that Council instruct the Space Group to begin a study of a European Space based defence system.[151] In a sense, this can be regarded almost as a political fall-back position - a recognition that while, in an ideal world, unfettered by budgetary and other constraints, the speedy implementation of a space-based system might be desirable, in the real world such a proposition may be politically unacceptable.

This moderation is also in evidence elsewhere in the report. In paragraph 80 of the explanatory memorandum it is pointed out that there is often a wide gulf between fine words and reality, citing the enthusiasm and commitment of the French in relation to military space, yet reminding the Assembly of the difficulty that has been encountered in France finding partners for its Helios 2 programme. It goes on to highlight the fact that in the absence of cooperation and the development of a space policy albeit minimal, efforts undertaken are likely to be affected by short-term considerations, basically of a financial order, and later warns that the challenges must be fully understood before proceeding further: 'The high cost of space does not allow wastage at a time when resources are increasingly scarce.'[152]

The Lisbon Ministerial Council had specifically tasked the Space Group to examine three options, namely the establishment of a WEU satellite system, participation in a developing multinational programme, or the continued procurement of imagery. The timescale given to the Group was that they should present a proposal to the Autumn 1995 Ministerial meeting. Each of these three options presented different possibilities. The development of an independent WEU satellite system would naturally provide the greatest operational possibilities, and allow the organisation total control over all phases of the gathering of intelligence. Against this was the cost. Such an option would be the most expensive, costing an estimated $10-11 billion over a ten year period. The second option, that of participating in an existing multinational programme would be cheaper, but would mean that the WEU did not have total control over the system. Interestingly, the programme is not identified by name, but can only refer to the Helios 2 programme, that being the only such multinational programme on the drawing board. The final option was effectively the status quo - that the Centre should continue, as it presently did with simply acquiring imagery from other sources and analysing it. This, the cheapest option - at least in terms of initial financial outlay - offered the most limited potential for the further development of the Centre's intelligence capability.

At the end of the day, when the Ministers met to consider the options, the report on 'The development of a European space based observation system'[153] is rather prescient. For indeed, the ultimate decision of the WEU Council was not to proceed with a full European space-based observation system which was considered to be too ambitious a programme to embark on from scratch. When they did finally come to a decision on this, at their meeting in Madrid, in November 1995, the final decision was that, at least for the time being, Europe's fledgling satellite intelligence capability would have to continue to rely on existing space-based resources for the data. This decision seemed to represent a half-hearted, lowest common denominator approach to an area of intelligence vital for European security. One is forced to assume that in the ultimate cost benefit analysis, the costs were simply too high for the perceived benefits.

Nevertheless, this is not the end of the story. While the independent, autonomous WEU satellite system was discarded at the Madrid meeting in November 1995 the Space Group was requested to review whether WEU should establish its own ground segment to receive and disseminate satellite imagery. Such a move would, to some extent, enhance the operational capability of the Centre, eliminating the sometimes lengthy delays in acquiring the desired imagery. Furthermore, while the option of becoming involved in a 'developing multilateral European programme' (presumably Helios 2) was not approved at that time, it was reported that the WEU was exploring what steps it must take in order to become a full partner in this programme.[154] The same press report also suggested that the WEU Ministers 'gave at least the appearance of reinforcing WEU's ties with Helios by naming Col. Bernard Molard, a Helios programme manager at the French Defence Ministry, as the WEU satellite centre's new director'[155] Col. Molard, who assumed his new role as the second director of the Centre in January 1996, was reported as having previously said that if more than four countries participated in the Helios 2 programme, a dedicated European organisation in charge of strategic space based observation would require to be created.[156] This creates an interesting possibility. If, in the future, the WEU were to go ahead and assume partnership status within the Helios 2 programme one wonders whether, in relation to Col. Molard's comment above, that would be construed as one more partner, or an automatic multiplicity of partners. Furthermore, it gives rise to the obvious question as to whether the WEU would be regarded as a suitable candidate in its own right to take on the role of that dedicated European organisation.

The possibility of WEU involvement in the Helios 2 programme (and also a proposed German led synthetic aperture radar satellite programme: Horus) was further strengthened with a very strong call for such involvement by the WEU Assembly, following a report by the Technological and Aerospace Committee on this topic which was discussed at the meeting of the Assembly in June 1996. The resulting Assembly recommendation, 595, calls upon the Council to:

> '... study arrangements for possible participation by WEU in the HELIOS 2 and Horus programmes taking account of the following:
>
> a) owing to the advanced stage of the Helios 2 programme it is inconceivable for WEU to play an industrial role in the programme;
> b) WEU participation in the programming of Helios 2 is both desirable and feasible;
> c) such participation in the satellite's programming must be based on WEU's strategic needs as an organisation and not on a mere aggregate of various national requirements;

d) it is still possible for WEU to participate in the HORUS programme in both an industrial and programming capacity.'[157]

In presenting the draft recommendation to the Assembly, Christian Lenzer the Rapporteur for the Technological and Aerospace Committee reiterated that while Europe had no independent space-based reconnaissance capability, governments need an independent system of this kind in order to use their own judgement to avoid crises wars and disasters. Indeed he went on to say that he was inclined to consider it almost as indispensable strategic information, and that it was absolutely essential for the member states of WEU to have an strategic worldwide, all-weather, space based observation capability to allow them to fulfil their political commitments in the western Alliance.[158]

The Recommendation also called for a number of other steps to be taken to improve the operational capability of the Satellite Centre, which will be further discussed in chapter seven, which examines in more detail the future of European military space policy.

V. Conclusions

The creation of the WEU Satellite Centre at Torrejon and its subsequent achievement of permanent status as a WEU body undoubtedly represents a significant step forward in the realm of European military space policy. It has not, however, been a straightforward, simple progression. The initial decision to establish the Centre in its experimental phase can be seen as a fairly brave decision on the part of the Ministerial Council albeit after many years of pressure from the Assembly. However it has not truly lived up to its original expectations. In its first three years of operation, the Centre completed 45 satellite image assignments on behalf of the WEU Planning Cell, or individual member states. This does not seem to be a particularly heavy workload. In truth, there seems to have been a reluctance to utilise the resources of the Centre, which has also, according to at least one report, received only tepid support from the Ministerial Council.[159]

The reluctance of the organisation as a whole to support the development of an autonomous WEU space-based segment of a satellite observation system, understandable, perhaps, due to the immense financial costs inherent in the development and operation of such a system, nevertheless would seem to limit the ultimate potential of the Satellite Centre as a wholly independent, autonomous European resource for intelligence analysis. One cannot help but think that if, during the three years in which the Centre has been in operation so far, it had been able to generate a greater output, in a more timely manner, the Council and individual member states might have been more convinced of the advantages to be gained from such a facility. In this regard, however, one must not be too critical of the Centre itself. Initially, of course, energies were devoted to its actual establishment, and the rapidity with which this was achieved is highly commendable, saying a great deal for the commitment and the enthusiasm of the original director and his staff. Furthermore, in the early days of the Centre's operations a considerable proportion of time had to be devoted to the training of staff. The Centre is also constrained, again for financial reasons, as to the quantity of imagery it can procure. Clearly, if it were to suddenly be faced with a significant upsurge in demand for its output, it may well face budgetary difficulties in attempting to satisfy the increased demand. Nonetheless, one is still left with a feeling that it is under-used.

Using some extremely simplistic arithmetic, if one accepts that it has undertaken 45 satellite image assignments in three years, this gives an arithmetic mean of 15 per year. The 1996 budget for the Centre is $13.8 million. If it were to complete 15 assignments in 1996, this would seem to indicate that the average cost per assignment is in the region of $920,000 per assignment, which seems rather on the expensive side. Clearly, such crude arithmetic is far from being an accurate representation of the actual cost of any given assignment. It takes no account of the other elements of the Satellite Centre's work, such as training etc. which is in itself a valuable role. In the absence, however, of a more detailed breakdown of the Centre's budget, it serves as a very rough indicator of the highly expensive nature of such an enterprise.

The initial difficulties which the Centre faced seem, to some extent to have been overcome. There is little doubt now that the Centre runs efficiently within the inevitable financial constraints. Issues such as the method of delivery of imagery are currently under investigation. Indeed, it is fair to say, that the Centre is still going through an on-going evaluation of the way in which it operates, with a view to further increasing its operational capabilities and the quality of service which it can offer to both the WEU as an organisation, its individual member states, and other potential clients. It is an evolving, dynamic entity which has got off to a reasonably good, albeit not faultless start. There is little doubt that as time goes on, its utility will be further recognised, its operational capability further improved, and its capacity to provide timely analysis of imagery required for the security of Europe greatly enhanced. Barry Blaydes, addressing a EUCOSAT symposium in 1995 described the future of the Centre as very promising, but dependent on other developments over the next few years. Internal considerations, such as the acquisition of high resolution imagery (such as that from Helios) and a quick and reliable response to users, he stated, must be complemented by other advances. He drew particular attention to the need for a radical improvement in the entire process of the acquisition of imagery, from whatever source.

> 'If the WEU has no medium term prospect (meaning in the next two to three years) of a satellite system under its direct control, then for the time being the Centre needs direct communication with and probably a presence at, the site of the satellite system control centre. There is also a need for the direct communication of imagery from the satellites to the Centre. These measures could easily be implemented over the next few years.' [160]

The other area which he identified as being of importance to improve was the manner in which tasks were formulated and transmitted to the Centre, together with the means to return the results to those who needed them. Clearly therefore there are a number of quite significant areas of the method of operation of the WEU Satellite Centre which still require to be examined and improvements incorporated into the way in which it carries out the tasks allotted to it.

However, the Satellite Centre, as a WEU body is here to stay. Arguably, it has done as much as could reasonably be expected of it in its initial phase of operation to prove its worth - at least in clearing the hurdle of achieving permanent status. In a sense, the future development of the Centre, the amount of work it undertakes, and indeed its effectiveness, rests in the hands of others. Fundamentally, two elements, external to the Centre itself, have to be enhanced for it to grow. The first thing that is required is an increasing appreciation of the value of this resource both by the WEU itself, and also by other users. The second, is the continued, or increased political will of the organisation and its various member states to use and support the Centre, backed up by the necessary financial commitments. Only if these two elements are present will the Satellite Centre be able to reveal its true potential as a significant component of the autonomous security infrastructure of Europe.

ENDNOTES

[123] Assembly of WEU, Eighth Ordinary Session (First Part), Report submitted on behalf of the Committee on Defence Questions and Armaments, *European Space Organisation*, Document 232, (Assembly of WEU, Paris, 3 May 1962)

[124] Van Eekelen, W., Speech at inauguration of Satellite Data Interpretation Centre, April 1993

[125] See for example Assembly of WEU, 36th Ordinary Session, (First Part), Report submitted on behalf of the Technological and Aerospace Committee, *Observation satellites - a European means of verifying disarmament - guidelines drawn from the symposium*, Document 1230, (Assembly of WEU, Paris, 25 May 1990) and also Assembly of WEU, 38th Ordinary Session (First Part), Report submitted on behalf of the Technological and Aerospace Committee, *The development of a European space-based observation system*, Document 1304, (Assembly of WEU, Paris, 30 April 1992)

[126] Van Eekelen, W., Speech at inauguration of Satellite Data Interpretation Centre, April 1993

[127] See the WEU Council of Ministers, Communique, (Vianden, Luxembourg, 27 June 1991) included as Appendix Three

[128] See Appendix Three

[129] See Appendix Three

[130] See Appendix Three

[131] WEU Satellite Centre, *Information Brochure*, (WEUSC, Torrejon, 1996), p.4

132 EUCOSAT, 'Towards a European satellite monitoring Agency', *Proceedings of Symposium*, (EUCOSAT, Paris, June 1993), p.89

133 EUCOSAT, 'Towards a European satellite monitoring Agency', *Proceedings of Symposium*, (EUCOSAT, Paris, June 1993), p.90

134 Blaydes, B., Official Record of Colloquy - a European space-based observation system held in Gran Canaria 24-25 March 1995, (Assembly of WEU, Paris, 1995), p.35

135 Blaydes, B., 'The Torrejon Centre and its Prospects', *Proceedings of the Second Eucosat Symposium, Satellite System for Security: A European Multi-User System*, 20/21 September 1995 (Eucosat, Paris, 1995)

136 Personal letter from Asbeck, F., Deputy director of the WEU Satellite Centre, 9 April 1997

137 Blaydes, B., *Official Record of Colloquy - a European space-based observation system*, held in Gran Canaria 24-25 March 1995, (Assembly of WEU, Paris, 1995), p.36

138 WEU Satellite Centre, *Information Brochure*, (WEUSC, Torrejon, 1996), p.6

139 WEU Satellite Centre, Ibid., pp.6-7

140 WEU Satellite Centre, Ibid., p.2

141 Assembly of WEU, 40th Ordinary Session, (Second Part), Report submitted on behalf of the Technological and Aerospace Committee, *The future of the WEU Satellite Centre in Torrejon*, Document 1437, (Assembly of WEU, Paris, 9 November 1994), p. 188

142 Assembly of WEU, 39th Ordinary Session, (Second Part), Report submitted on behalf of the Technological and Aerospace Committee of the, *The development of a European space-basedobservation system - Part II*, Document 1393, (Assembly of WEU, Paris, 8 November 1993), p. 174

143 Assembly of WEU, 41st Ordinary Session, (Third Part), Report submitted on behalf of the Defence Committee, *A European Intelligence Policy,* Document 1517, (Assembly of WEU, Paris, May 1996)

144 *Space News*, Vol.6, No.45, Nov 27-Dec 3, 1995, p.21

145 Hayward, K., *British Military Space Programmes*, (Whitehall Paper, RUSI, London 1996), p.61

146 WEU Council of Ministers, *Communique*, (Lisbon, 15 May 1995)

147 WEU Council of Ministers, *Communique*, (14 October 1991)

148 Assembly of WEU, 40th Ordinary Session, (Second Part), Report submitted on behalf of the Technological and Aerospace Committee, *The development of a European space-based observation system - Part III,* Document 1436, (Assembly of WEU, Paris, 9 November 1994), p.179

149 Assembly of WEU, Document 1436, p.179

150 Assembly of WEU, Document 1436, p.181

151 Assembly of WEU, Document 1436, p.173

152 Assembly of WEU, Document 1436, p.181

153 Assembly of WEU, Document 1436

154 *Space News*, Vol.6, No 45, Nov 27 - Dec 3, 1995, p.21

155 *Space News*, Ibid., p.21

[156] *Space News*, Ibid., p.21

[157] Assembly of WEU, Recommendation 595 on. *WEU and Helios 2*, Proceedings of the 41st Ordinary Session (Second Part), (WEU, Paris, June 1996), p.33

[158] Assembly of WEU, Proceedings of the 41st Ordinary Session (Second Part), (WEU, Paris, June 1996), pp.106-7

[159] *Space News*, Vol. 7, No. 22, June 3-10, 1996, p.21

[160] Blaydes, B., *Proceedings of 2nd EUCOSAT symposium*, held in Bonn-Petersberg, 20-21 September 1995, (EUCOSAT, Paris, 1995), p.145

CHAPTER FIVE

MILITARY SPACE AND WESTERN EUROPE'S CURRENT CAPABILITIES

I. European Space Defence

For many years transatlantic consideration of security and defence, particularly within the NATO framework, has been coloured by the underlying concept of burden-sharing. At times a source of significant tension within the North Atlantic alliance, there has been considerable debate as to the extent to which Europe should bear the responsibility for its own defence and security.

Traditionally, this debate has been most vocally carried out at times of change in global security perceptions. In the latter part of the 1980s, and the 1990s, as the Cold War drew to a close and states throughout the world struggled to come to terms with a vastly changed international arena, while simultaneously facing economic difficulties which were giving rise to vigorous demands for a reduction in defence expenditure, such a debate was as topical as ever.

However, up to that time, there was one specific area of defence expenditure that hitherto had been excluded from this transatlantic discourse and debate: the utilisation of space as a military asset. This element was excluded by omission rather than through deliberate policy, simply because it was an area in which the US had traditionally been seen to have by far the greatest resources, a pattern which her European allies had shown comparatively little interest in altering for the simple reason that there seemed little point in wasting scarce resources by duplicating facilities on a function for which the Europeans were not convinced there was a pressing need. The result of this situation in terms of Western security, is that the US has unquestionably been the single most important actor in the military use of outer space. It is, of course, true that both France and the United Kingdom have, for a number of years, operated national military communication satellite systems, as has NATO itself. However, until the 1990s that was virtually the extent of European involvement in military space activity, something which was in no way reflective of the United States penetration into all aspects of the military applications of space.

The last few years however, have seen this situation slowly begin to change. For the first time since the end of World War II very serious attention was paid by a number of states, to the possibility of establishing a multi-national military space capability within Western Europe. A number of parallel suggestions appeared on the political agenda for consideration by different combinations of countries in Europe. Perhaps the most potentially significant in terms of long-term planning was that the Foreign and Defence Ministers of the Western European Union States began to seriously consider their future security requirements in terms of surveillance satellites, throwing up the very real possibility that they would agree to establish a regional satellite monitoring agency with its own space-based component, in essence a Western European military satellite system.

Such a policy decision would represent a major step towards the 'Europeanisation' of security and defence policy.

II. Background

In order to fully comprehend the current situation regarding the possible establishment of a European military surveillance satellite system it is necessary to first consider why such a system has not, until now, been regarded as desirable. Furthermore, given the increasing number of international organisations professing a stake in European security, why has the role of investigating such a step been assumed by the Western European Union, rather than one of the more historically visible and active international organisations in Europe, such as NATO?

The use of space for military purposes is not a new phenomenon. Indeed this concept existed before any of the technological capabilities which might make it a reality had been developed. Throughout the earliest days of space research in the 1950s almost all such work was carried out under the auspices of the military in both the United States and the Soviet Union. However, the European experience differed somewhat, with much of the impetus behind space research being provided by civilian scientists. One example of this is the attitude of one of the great pioneering space scientists, Eduardo Amaldi, whose original proposal for a European organisation for space research was 'based on one main pillar: the military were to be kept out of the future European space organisation'.[161] Although this view was not shared by some of his peers, it was to become the prevailing opinion of the majority of those scientists involved.[162] Thus, from the earliest days, the Europeans did not regard the military aspect of this technological development with the same degree of significance as the Americans.

The launch of the first Soviet Sputnik satellites in 1957, and the series of launches which followed on from that first example of a man made satellite orbiting earth, not only provided the first strike in the Cold War 'Space Race', but also provided an excellent example of the differing perspectives such a development generated on opposite sides of the Atlantic. Although both sides of the NATO alliance were aware of the strategic implications of such a capability, it was the Americans who regarded it as the greater threat.

This is illustrated by the response of the countries politicians. In the US, the House of Representatives adopted Resolution 1770, in May 1958, just after the launch of Sputnik 3. This document stated that:

> '... this country is not unmindful of what these Soviet achievements mean in terms of military defense ... The United States must have strong capability in the use of outer space, both as a deterrent to the use

of military vehicles against this country and as an aid in developing antimissile techniques.'[163]

In Europe, however, the successful Soviet spacc launches engendered a somewhat different response. In the UK for example, Denis Healey MP, in an address to the Royal Institute of International Affairs, was quick to point out that:

> 'Thus the Sputnik has not only changed the global balance of power in favour of the Soviet Union, it has also changed the balance of power inside the alliance in favour of Europe. America now believes her own survival depends on obtaining new facilities from her allies. And she is simultaneously asking her allies to base their survival on a strategy which would spare the United States from direct attack. The allies are in a position to demand a very high price from the United States for accepting these new risks ... The sputnik itself has not increased the vulnerability of America's allies, which was already absolute in total war. But the American reaction to the sputnik may transform the whole character of the alliance.'[164]

This difference in reaction is logical. To the Americans, bounded by two oceans, the development by the Soviets of a technological capability to utilise rockets provided a significant alteration in their perception of the threat posed by the Soviet Union. Whereas prior to this the mainland of the United States had been virtually invulnerable, the capability to launch rockets on a ballistic trajectory was seen as evidence enough that the US was now much more open to a decisive strategic attack. Having thus rendered the United States vulnerable, a hitherto unprecedented circumstance, the attitude was perceptibly different. This contrasts with Europe which, as part of the same land mass as the Soviet Union and its Eastern bloc satellite states, still perceived the main threat as that of a conventional military assault. It is therefore not surprising that amongst the countries of the West it was the Americans who chose to undertake more serious development of missile and rocket technology in a bid to counter the strategic initiative that it was felt Sputnik had wrested from them.

It is a logical progression that the US should see space as another arena in which they might attempt to achieve Cold War foreign policy gains. Through the space race which was an imposing feature of superpower relations throughout the 1960s, the United States, ever-mindful of the possible military advantage that could accrue from the pursuit of space technology, vigorously applied a policy designed to develop their advanced military space capabilities.

In the European context, however, such a route was less necessary. In the first instance the direct threat was of smaller proportion. Secondly, within the overall framework of NATO there seemed no need to duplicate such efforts. Since the US was keen to move forward in such areas, there was much less need for Europe to go down a similar path, indeed it would appear that this was an area in which the United States was relatively content to allow its European partners to 'free ride'.

So the responsibility for, and investment in, Western military space policy was largely defined by the security priorities dictated by geography. Europe did not, of course, completely ignore this area, with both the British and the French developing national military communication satellite systems in the shape of Skynet and Syracuse respectively. Yet this was largely the limit of their commitment to the military use of space and represented the introduction of one very specific system to fill a particular national requirement. Only the Americans developed the panoply of space-based military assets seen today, with satellite systems for tasks as diverse as early warning, reconnaissance, surveillance and communications, demonstrating their commitment to the broad range of activities in this field of endeavour.

With these points firmly in mind, It is not difficult to understand why, within the Western alliance, the US should be pre-eminent in the utilisation of military space systems. Indeed the more interesting question is why, at this particular time, Europe should be considering enhancing its own capabilities in this area?

III. Changing Perceptions of European Security

Throughout most of the last half-century, there has been an unwritten belief that the security requirements of 'the West' are synonymous with those of the United States. Such a situation has persisted from the earliest days of the post-World War II alliance. The American commitment to European security through the over-arching framework of NATO has largely remained constant. Nevertheless, within that framework there has, from time to time, been a certain tension, with periodic assessments looking into issues of: 'burden-sharing' - the extent to which Europe should assume responsibility for her own security; and the continuing need for an American military presence in Europe.

The dramatic changes in the European security environment during the 1990s, have precipitated an intense, ongoing re-evaluation of European security requirements. Difficult questions about the need to retain the status quo have been, and continue to be asked. Each of the main bodies with a security responsibility, such as NATO, OSCE, WEU, and now even the European Union is trying to establish a new role for itself in the changed circumstances. Within the US the question has been asked as to whether, in a period of financial restraint, a costly US presence in Europe is still required.

The interim answer to that particular question remains a fairly clear affirmative. President Clinton, for example, addressing the North Atlantic Council Summit in Brussels in January 1994 opened his remarks by saying that 'the security of the North Atlantic region is vital to the security of the United States.'[165] Be that as it may, it does not mean that roles and relationships will or should remain unaltered.

The new transatlantic security relationship is neatly defined by Asmus, Kugler and Larrabee, who define six steps which are necessary in forging a new transatlantic bargain.[166] The first of these steps clearly identifies the changed circumstances now prevailing:

> 'While there remains a general consensus in the United States that Washington should remain engaged in Europe, this no longer automatically translates into support for the old transatlantic bargain.

> Changing American priorities mean that the United States now needs to share the burdens and responsibilities with its allies. European political imperatives have also changed in ways that mirror the United States. An overwhelming desire to see the Americans remain engaged in Europe notwithstanding, Europeans want a more balanced and equal relationship with Washington.' [167]

It is therefore clear that the European side of the partnership is destined to become more active in areas hitherto largely left to the Americans. Evidence for such changes can be seen in the growing range of activities of the Western European Union. Indeed following the January 1994 North Atlantic Council Summit, Sir Dudley Smith, President of the WEU Assembly was prompted to acknowledge the growing status of the WEU by stating that, 'For the first time in 40 years the WEU has been formally recognised as a rightful component of the transatlantic partnership - as complementary to NATO rather than as in competition.'[168] Further evidence of this enhancement of the European side of the transatlantic security system lies in the joint declaration issued at the conclusion of the North Atlantic Council summit which, in its first nine paragraphs, uses the phrase 'European security and defence identity' six times, mentions 'the European pillar of the alliance' five times, and speaks of the WEU no less than eight times.

As the new security architecture for Europe is taking shape, a number of concrete examples of the 'Europeanisation' of defence and security are also emerging, amongst them the introduction of WEU sponsored work in the field of military space activity. This specific development is of particular interest, representing a deliberate move into practical European cooperation in a field which previously was very largely the domain of the United States. The success or otherwise of this programme may not only be seen as a marker of the political will to develop independent, multinational European defence capabilities, but also as highlighting some of the practical difficulties inherent in such a venture.

However, it would be incorrect to assume that the European states have no interest in the military use of space. At present, in terms of military space capability, Europe lies third in the world, behind only the USA and Russia.

Why is it that space is now so important in military terms and what exactly are the current capabilities of Europe ?

It is difficult today to envisage any major power which does not rely to some extent on space resources for military planning and operations. In many militarily important areas, such as surveillance and navigation, space resources are today the most effective and efficient means of carrying out these crucial tasks. In other areas, such as ballistic missile warning systems and real-time global communications capability, space-based assets are the only means by which these facilities can be made available in an effective manner. Because of this increasing dependence on space-based resources, it is difficult to over-emphasise the significance of space to the modern military.

However, one should also bear in mind that military planners are not particularly interested in space *per se*. This point was well made by the UK MOD in 1987 when they stated:

> 'From the MOD's standpoint, space activity is not an end in itself and the Department does not have an overall space programme. Rather, space-based methods, where they are available as an option, are seen as a potential means of achieving given ends, to be assessed alongside other means of achieving these ends against criteria such as effectiveness and affordability. The high cost of space activities means that space is only employed when the benefit, in defence terms, is judged to be commensurate with the outlay.' [169]

Although this statement originates from the United Kingdom Ministry of Defence, it neatly encapsulates the attitude of almost all states in relation to the military use of space.

Military use of space may take two forms. In the first case, space is used to support more conventional planning and operations on earth. Secondly, space itself can be an area for direct military activity. These two forms correspond closely to the two traditional descriptions of space-based military resources: passive and active. A passive system is not in itself a weapon, but is used to support military activity, for example, surveillance or communications satellites. An active system is either a weapon in its own right, or an inherent part of an overall weapon system, such as space-based lasers or targeting systems. Both these areas of military activity centre on the use of satellites, whereas such items as space planes and the like remain in the realms of science fiction and may or may not be developed in the future. To date, no active military space system has been operationally deployed.

IV. The Military Uses of Space

At present, there are really three principal tasks of military value which are currently undertaken with the aid of satellites in earth orbit. These support the essential elements of command, control, communication and information (C^3I) which lie at the very heart of the modern military. The tasks specifically referred to are:

- communications
- reconnaissance/surveillance
- navigation

Communications is one of the oldest and most established functions of satellites, and remains in military terms, one of the most significant. The use of satellite communications (SATCOM) provides the ability to communicate with relative ease on a global scale, something which compares favourably with alternative systems. Of particular value is the use of satellites in Geostationary Earth Orbit (GEO),[170] an arena which, if utilised properly, requires a constellation of only three satellites to give virtually global coverage (with the exception of the polar areas). Accordingly, such an orbit is highly suitable, and is widely used, for communications purposes, not only by the military but also by commercial enterprises.

Military communications via satellite may be part of a dedicated military payload, solely for use by the military, or alternatively can be part of a military package on a dual use civil/military satellite. It is in the area of communications satellites that the greatest multiplicity can be seen. The US, for example, have several different military communications satellite systems. These include the Defense Satellite Communications System (DSCS), which provides a secure facility for commanders to talk to each other, and which can link with every major command unit of the widely dispersed US forces.[171] Furthermore, such systems as FLTSATCOM, which offers satellite communications to the Navy, Air Force, and National Command Authority, and the separate Air Force AFSATCOM, all add to the spread of space based communications capability. Indeed it is estimated that some 80 per cent of US military communication is now carried via satellite.[172]

Reconnaissance and Surveillance is another major role which is undertaken by satellite. Of course the terms 'reconnaissance' and 'surveillance' are not strictly speaking synonymous with regard to the technical collection of intelligence. Taken in isolation 'reconnaissance' refers to the active search for particular data (e.g. the monitoring of radar transmissions to determine the operational capabilities of the originating system) whereas 'surveillance' refers to the more general, passive watching or listening for something to occur (e.g. the activation of a radar system or the launch of a missile). However, for the sake of linguistic simplicity, in this chapter, all such monitoring activity shall be referred to as 'reconnaissance'.

Reconnaissance, in one form or another, has traditionally been a key element of military planning. Knowledge of the disposition of enemy forces thus gained is perhaps the most important intelligence sought by a military commander. From the earliest days of scouts on hill-tops, through the spotting balloons of World War One and the U-2 flights of the 1960s, new means of obtaining such intelligence have constantly been sought. The technical advance of satellites has been able to offer a vastly enhanced observational capability and thus represented a significant breakthrough in the reconnaissance field.

The use of satellites to 'observe', from space, military activity on Earth, is perhaps one of the most useful, and certainly one of the longest established programmes in the space-assets arena. The first active photo-reconnaissance satellite was the American TIROS 1, launched in April 1960.[173]

Essentially, there are four distinct types of reconnaissance satellites: imaging, or photo-reconnaissance; electronic reconnaissance; ocean reconnaissance; and early warning satellites.

Imaging satellites are the commonest form, accounting for around 80 per cent of all reconnaissance satellites launched.[174] Such satellites use photographic and television cameras, multi-spectral scanners, infra-red sensors and microwave radar to observe the earth.

These undertake two distinct types of mission: area reconnaissance, in which relatively large areas of the planet's surface are scanned to allow observation of, for example, large scale troop movements, or the construction of new buildings in a militarily sensitive area; and close look, high resolution reconnaissance which permits the detailed observation of a small area. Obviously, very detailed examination of large areas is, if not inherently undesirable, at least utterly impractical, for two reasons. First, the technical capacity to undertake such a task would require hundreds, if not thousands, of satellites which would be a prohibitively expensive undertaking, and secondly, the sheer volume of data collected would completely swamp the relevant intelligence agencies. Thus we find that these two distinct mission profiles have developed and in the main, the two missions are undertaken by separate satellites. While it is possible for one satellite to be capable of fulfilling either role, as is the case, for example, with the American KH-12 satellite, this multi-role capability remains unusual because of the quite different nature of the orbits required for optimum results. Area reconnaissance is best carried out from a much higher orbit than is viable for close look reconnaissance, and the ability of a satellite to change orbit is constrained by its fuel and booster capacity. It is still generally easier and cheaper to use different satellites for the different roles.

The ability of an imaging satellite to 'see' objects is determined by its ground resolution, which can be defined as the size of the smallest object on the ground that can be distinguished with good contrast.[175] Currently, military satellites using photography can achieve a ground resolution of around 10-30 cm, making it possible to identify targets including artillery and small vehicles.[176] It is therefore quite possible to distinguish between, for example, a tank and a lorry.

While actual photographs are the easiest form of data for the layman to interpret, the other forms of imagery which are currently available are no less useful to the experts. The use of multi-spectral images gives much more information because the chemical properties of different substances cause them to take on different appearances in various spectral bands, thus showing the distinction, for example, between real vegetation and camouflage.[177]

Similarly, differences in the heat of objects is apparent if infra-red sensors are used. While it is possible to camouflage or hide a tank, for example, it is more difficult to conceal the infra-red 'signature' of its engines. Indeed, such infra-red scanning can reveal targets which have moved and are no longer there:

> 'Airplanes, for example, cause shadows on the concrete aprons where they are parked even in hazy weather. Hours after the plane has departed, its shadow silhouette, which is cooler than the surrounding area, remains visible to the infra-red sensor.'[178]

However, all these means of observation have one significant drawback in that they depend to some extent on the weather conditions. Heavy cloud cover, something not uncommon in Central Europe, will hinder or defeat them. This problem is resolved by using microwave radar reconnaissance, which can penetrate clouds and needs no light. However, resolution remains a problem with this particular form of imaging, and a great deal of power is required to produce the microwave pulses. Such a system has long been operated by the Soviet Union and civilian or military radar satellites are planned by the European Space Agency, the United State, China, and Canada.[179]

The use of satellites for navigation purposes allows for a great degree of accuracy in pin-pointing the exact location of men and materiel. Both the Russian and American systems work on the same principles, and are actually very similar, although somewhat fewer details are known of the Russian system.

The US system is known as NAVSTAR (Navigation System using Timing and Ranging) or GPS (Global Positioning System), and uses a total constellation of 24 satellites.[180] Users of the system range from individual foot soldiers, carrying the necessary receiver which nowadays may be no larger than a pocket calculator, to ships, aircraft, or even ballistic missiles. The equipment necessary to use the system comprises a radio receiver and computer. This locks on to signals from the nearest GPS satellites and translates the data into a very precise navigational fix. The advantages of this system are its great accuracy (calculated to be around 18m),[181] its immediate worldwide coverage, and the fact that individual users are electronically passive and so cannot give away their position merely by using GPS. The Gulf War helped to prove the value of this system, with it being utilised for navigation and targeting purposes by all branches of the Allied forces. This value was noted by Paul Kaminski , US under-secretary for acquisition and technology in a statement to the National Security sub-committee of the House Appropriations Committee in March 1995:

> 'GPS proved of enormous military war-fighting value in Desert Storm as well as in subsequent military, humanitarian and support actions in Somalia, Bosnia and Haiti. In Desert Storm, GPS was largely responsible for the swiftness and precision of our force movements and for the relative safety with which they were conducted.' [182]

The Russian equivalent, GLONASS, is in many ways virtually identical to NAVSTAR/GPS, and is understood to operate in the same way. In addition to these military-based satellite navigation systems of the US and Russia, both West Germany and the European Space Agency have reportedly studied commercial systems.[183]

Electronic Reconnaissance satellites, more commonly known as Electronic Intelligence (ELINT) satellites, are designed to detect and monitor electronic emissions, such as radio or radar signals, emanating from the opposition's military equipment. Also known as 'ferrets' from their task of 'ferreting out' information, the US began with the Rhyolite series in the 1970s, around the same time that the USSR commenced its own programme of ELINT satellites (e.g. Cosmos 389).[184] The United Kingdom and France are also known to have launched, or planned to launch satellites with an ELINT capability, as discussed later in this chapter.

A brief but useful description of the role of ELINT satellites is given in Jane's Spaceflight Directory 1988-89:

> 'They are electronic ears, recording radio and radar transmissions from areas of military activity. When replayed to ground stations the 'radar signatures' - characteristics such as pulse repetition frequency, pulse-width, transmitter frequency and modulation - enable the likely function and method of operation of a particular centre to be identified. The number and type of electronic systems at a particular site, and subsequent changes in the signals, give a valuable indication of its purpose and capability.' [185]

Of course, in addition to information gleaned from the technical 'signature' of the signal which may reveal information such as the type of transmitter being used, it is also possible to make more direct use of transmissions which are being monitored, intercepting radio communications, for example. While certain techniques such as frequency hopping (the rapid changing of the transmission frequency) can be adopted to prevent accurate interception, the very fact that transmission is occurring at all can be of military significance to an opponent.

Like photo-reconnaissance satellites, ELINT satellites are of two types, one for large-area surveillance, normally at an orbital altitude of around 1500 km, with others, orbiting at altitudes which vary between 500 and 700 km, which are designed to obtain more detailed information.

Two rather more specific types of reconnaissance satellites are those dedicated to the roles of Ocean Reconnaissance and Early Warning.

As the name indicates, Ocean Reconnaissance satellites are used to observe the world's seas. First launched by the Soviet Union over 20 years ago to monitor US naval activity near the USSR, the scope and capabilities of such systems have increased steadily since then. Such satellites were regarded as more significant by the Soviets, and as a result were accorded a relatively high priority by them. This was in large part due to the greater naval capabilities of the United States and her allies in NATO. This ocean reconnaissance was regarded as a significant threat by US defence planners to the extent that much Anti-satellite (ASAT) research has been undertaken by the Americans. The significance of ocean reconnaissance is such that the vast oceans can be monitored and it is believed that the Russian satellites can relay, in real time, the position, heading and identity of naval forces.[186]

Russian ocean reconnaissance satellites fall into two categories: RORSATs (Radar Ocean Reconnaissance Satellites) and EORSATs (ELINT Ocean Reconnaissance Satellites). The former use active radar systems to carry out their task, while the latter obtain electronic intelligence from the target ships. Working in pairs, any attempt to counter the former by Electronic Counter Measures (ECM) will result in giving away the information, which it is sought to conceal, to the accompanying EORSAT.

The US system NOSS (Navy Ocean Surveillance Satellites, also known by the code name 'Project Whitecloud'), was first launched in 1976 to counter the increase in Soviet naval power. NOSS is believed to be a passive radar satellite, possibly incorporating infra-red sensors to track submerged nuclear submarines.[187]

In addition to their primary task of tracking vessels, ocean reconnaissance satellites are also used to monitor sea conditions. Such a capability improves weather forecasting and understanding of the oceans, which aids in the increasingly difficult task of submarine detection.[188]

Early Warning Satellites represent the first line of defence, as it is almost certain that the first indication of a nuclear ballistic missile attack on either superpower would originate from such space-based sensors. Both the US and Russia utilise early warning satellites to detect the launch of missiles. Their greatest value is in the increased time afforded to take appropriate defensive action. The use of satellite detection equipment, capable of spotting a launch within seconds of the missile being fired, allows the earliest possible alert, and as a result increases the warning of a missile attack. As an example it would increase the warning time of a missile launched between the mainland of the US and Russia to something in the region of 25-30 minutes.[189]

Although the early warning satellites would most probably give the first indication of a nuclear attack on either superpower, no response would be based on such data alone. Indeed in the course of any year, each side observes many space launches and missile test firings. The threat of attack must be confirmed by another source, normally ground-based early warning radar. Such a procedure is known in the West as 'dual phenomenology'.[190]

V. Current European Military Space Capability

The following sections trace the various involvements of European states in the exploitation of space based technologies, with the type of system used as the criterion of differentiation.

Communications Satellites

At present the UK, France and Spain have operational communication satellite programmes. In addition to these national efforts, there are a number of other programmes which are currently at the planning or development stage and involve a number of other West European nations. Indeed, in the sense that NATO has its own communication satellite system, it can, rather more tenuously, be argued that all the European NATO member states have a 'share' in a military satellite communication system, although it is not really possible for them to be construed as independent military satellite operators.

The currently operational communication satellite systems in Europe are the UK's Skynet 4 series, France's Syracuse 1 and 2, Spain's Hispasat 1A and 1B and the NATO 4 series.

Skynet

With Skynet 1A, launched in November 1969, the UK became the first nation in the world to successfully develop and deploy a military communications satellite into Geostationary Earth Orbit. This satellite provided a communications facility between the UK and the Middle and Far East for some five years.[191] Since then, the Skynet series has been continually redeveloped and updated. Today, the UK operates the fourth generation of Skynet in its Skynet 4.

The Skynet 4 system, which was initiated in 1981 aimed to develop satellites which would offer a specific range of services, including: an ability to support maritime operations spread over substantial areas of the earth's surface; provision of strategic/trunk links using relatively large terminals; a service to mobile terminals, mainly in the West European area; and strong system resistance to all forms of electronic warfare and other military threats. These objectives were to be achieved with the minimum of interventionist system control, and the satellites should have the maximum capacity to continue to offer these services in the event of partial failure, or even enemy attack.[192]

Accordingly, to accommodate these requirements, the Skynet 4 satellite incorporates two UHF transponders, principally for communication with submarines, an SHF package with four channels which is used principally for communication to and from ground stations and surface vessels, and one EHF uplink channel for propagation experiments for future EHF systems. Allied to these transponders are a range of antennae, permitting a number of options in terms of the receiving footprint - allowing earth cover, wide beam, narrow beam or spot beam. For security, and in a bid to counter electronic warfare, the satellite carries signal processing and anti-jamming equipment, and is hardened against electromagnetic pulse. In addition to the communications traffic, the payload is also used to carry telecommands and telemetry data.

In an effort to ensure that the system will continue to fulfil its intended role reliably throughout the designed seven year operational life of the satellite, many of the payload units are duplicated on board, so providing a back-up in the event of the primary unit failing, or being damaged.

The first Skynet 4 satellite was due to be launched in June 1986 by the US space shuttle. However, the Challenger accident in January of that year, which grounded the shuttle fleet for 32 months, meant that alternative launch facilities had to be obtained. As a result of these delays, the first in the series to be launched into orbit, in December 1988, by an Ariane 4 launch vehicle, was in fact Skynet 4B, which was declared operational on February 22 1989.

Skynet 4A was launched almost exactly one year after 4B, in December 1989 from Cape Canaveral, while the final satellite in phase one of the series, Skynet 4C, was launched on August 31 1990.

From the outset, it was envisaged that a second phase of the Skynet 4 programme might be implemented, to extend the life of the programme, with further, virtually identical satellites being launched at a later stage. This option was incorporated into the original plans for the Skynet 4 series to ensure continuation of service if no decision on successor system had been taken before the projected end of the active, working lifespan of the initial satellites in the series. It is, of course, impossible at the outset of a programme to predict accurately the effective lifespan of a satellite. This is contingent on, for example, the accuracy of orbital insertion at the time of launch, and thus the amount of on-board fuel used for station-keeping on orbit, the rate of degradation of batteries and other equipment etc.

In February 1992, with a view to the requirements of developing a second-phase Skynet 4 satellite, the UK government announced a £10 million study into possible enhancements to the original satellites, which might be incorporated into second phase satellites, if they were required.

Throughout the first half of the 1990s, considerable debate has been taking place on the nature of a successor system to Skynet 4. A number of options have been under consideration, which will be detailed later in this chapter. However, since no final decision on this matter had been taken by the UK government by 1994, it was duly decided to press ahead with the option of ordering three second-phase Skynet 4 satellites, to extend the operational capability of this system into the 21st century.

Thus, in 1994 two Stage 2 Skynet 4 satellites were ordered by the UK government, with a further contract for a third being awarded in February 1996. These three new satellites will replace the present SKYNET 4 Stage 1 constellation to meet the demand for increased capacity and coverage. In the annual Statement on the Defence Estimates (1996) it was announced that initial studies on a further generation of satellites (SKYNET 5) for introduction into service in the next century were completed in December 1995, and a range of national and collaborative options were being examined for the next phase of project definition.[193]

These three stage two satellites will replace the existing Skynet 4 satellites in the same orbital positions, with the first entering service in 1998 and the final one around the year 2000. A number of improvements in the satellites have been incorporated, principally in relation to increasing the available power of the communications channels, and allowing more flexible coverage. The first two of the three replacement satellites, Skynet 4D and 4E are scheduled for launch in May and December 1998, one by the McDonnell Douglas Delta 2 launch vehicle, and the other by Ariane 4. The final satellite, Skynet 4F is provisionally scheduled for launch in the year 2000.[194]

NATO 4

The NATO 4 system is the latest in a series of NATO communication satellites which were first mooted in 1966, when the NATO Council approved in principle the establishment of a NATO-wide communications scheme, and agreed to study whether a NATO satellite communications programme should be established as a cooperative effort by member nations.

The outcome of this study was a decision to proceed, and the first NATO communications satellite designated NATO 2A, was launched, in March 1970, from Cape Kennedy.

The current series of NATO satellites, NATO 4, is a version of the UK's Skynet 4 (chosen in 1987 over the rival US DSCS 3), and thus is essentially similar to the system already outlined. In fact, while owned by NATO, the satellites are operated on the organisation's behalf by the British Ministry of Defence.[195] Again, as was the case with the Skynet 4 programme, it was originally envisaged that the NATO 4 satellites would be launched by the US space shuttle, but in view of the delays this programme suffered, the first, NATO 4A, was placed in a geosynchronous transfer orbit by a McDonnell Douglas Delta 2 expendable launch vehicle from Cape Canaveral on 7 January 1991.

However, doubts have been raised as to whether the NATO alliance will continue to operate their own, independent communications satellites in the future. The NATO 4 series is due to reach the end of its effective lifespan around the turn of the century. However, at the end of 1994 Air Commodore Gabby Ferenczy, director of the command, control and communications directorate at NATO Headquarters, said that there was virtually no alliance thinking on policy related to satellite communications post-2000. Earlier in the year, addressing a military space conference, he had said he favoured continued NATO ownership of spacecraft, but 'short-term bureaucratic policies' within the alliance would likely overrule that view and opt to lease transponders aboard civilian satellites instead.[196]

Syracuse

The Syracuse system is a satellite based secure communications system for the French government and for French military forces- indeed the name is a French contraction of 'satellite radio communications system' (Systeme de Radio-Communication Utilisant un Satellite). Unlike the British and NATO systems, it is not based on independent single-mission dedicated satellites, but rather is incorporated into the design of the essentially civilian Telecom 1 and Telecom 2 multi-mission satellites. The decision to implement the Syracuse programme was taken in January 1980, and the first Telecom 1 satellite, with its military Syracuse payload on board was successfully launched into geostationary earth orbit in August 1984. This was followed by TELECOM 1B in May 1985 and by 1C in March 1988 (the originally scheduled launch date was June 1986, but there were delays with the Ariane launch vehicle).

The Syracuse system involves the use of two transponders on each of the Telecom 1 satellites, and provides round-the-clock secure communications facilities. In addition to providing a means of voice communications, the Syracuse system also supports telegraph, data transmission and fax services. In order to ensure the greatest possible security, messages can be enciphered and additional electronic methods are also built into the system to minimise the risk of interception or jamming.[197]

In 1987 the go-ahead was given for the design and development of the Syracuse 2 successor system programme, for which a total of some FF8 billion were budgeted. The Telecom 2 satellite is larger and more efficient than its predecessors, and has a projected lifespan of 10 years, rather than the 7 years offered by its predecessor.

The current military element of the system, Syracuse 2, offers several distinct advantages over previous systems. It has facilitated a large increase in the number of links, and also in the number of ground stations within the system, which can be lighter, and make use of the system's facilities by using a smaller antenna - which will permit the terminals to be incorporated into aircraft and submarines; allow inter-operability with other fixed and tactical communication networks; make use of coverage zones which are better suited to French requirements in the 1990s. Just how broad the communication network provided by Syracuse is can be evidenced when we consider the extent of French global interests:

> 'The Syracuse 2 system is aimed at providing all three French armed forces with permanent and secure links between government and defence authorities in France and troops deployed abroad. This includes, the Mediterranean and Middle East, the Atlantic and Indian oceans, and overseas territories in North and South America and the French Caribbean.' [198]

The third Telecom 2 satellite (2C0 was launched in December 1995, joining Telecom 2A and 2B in orbit. A fourth (2D) was scheduled for launch in 1996.

Hispasat

Hispasat is a Spanish communication satellite programme, which, by launching Spain's first telecommunications satellite, offers communication facilities to both civilian and military users. The programme as a whole comprises a total of three satellites, two operational and one spare, and the necessary ground control stations to command, control and track the satellite system, as well as the communications facilities themselves.

The first satellite (1A) was launched in September 1992, with the second (1B) going up in June 1993. Both satellites are Geostationary Earth Orbit at approximately 31 degrees West, within about 70 km of each other. One potential difficulty with this is that this segment of GEO is becoming comparatively crowded, with the inherent risk of interference between adjacent satellites which that brings, and so considerable effort is going into ensuring there is no such interference between the HISPASAT satellites and others.[199]

Each satellite carries a total of 18 communications transponders. Two of the transponders are dedicated military communications transponders, another three are for other government communications requirements, three for radio and television, while the remainder are for other services, including telephone and commercial business communications. The coverage of the satellites includes the whole of Spain itself, with broadcast coverage of most of Latin America and part of North America.

The Hispasat programme was originally designed to meet an urgent need by Spain's military forces for such a communications capability, in particular for the Spanish navy.

Reconnaissance Satellites

Helios

The Helios system is a satellite observation system jointly developed by France, Italy and Spain. It was originally conceived and initiated by the French, who were adamant that they required an independent observation satellite system. This feeling stemmed in part from a general reluctance to remain dependent on American satellites to satisfy their military intelligence requirements, allied to the feeling that possession of an independent nuclear deterrent should be matched by an equally independent military satellite observation system. From this belief emerged the Helios programme, which was started by the French in February 1986. It was then opened up to international collaboration and both Italy and Spain joined the project contributing 14 per cent and 7 per cent of the total cost respectively, a figure which is estimated at some FFr 7 billion.[200]

The Helios system is to a very great extent based on the established and proven SPOT civilian remote sensing satellite system which the French have developed. The Helios platform, recorders and a great deal of the electronics involved will be the same as in the SPOT 4 satellite. The principal differences lie in the photographic equipment, and the protection offered to the telemetry against possible interference. Overall, the Helios system includes optical and infra-red surveillance capability with a ground resolution approaching 1 metre - a considerable improvement on the 10 metre resolution capable with current SPOT satellites.

The first Helios satellite (1A) was launched on 7 July 1995, by an Ariane 4 rocket, with a second Helios 1B under construction and scheduled for launch, if needed in late 1996. It proved very successful in an experimental programme designed to produce satellite imagery from Bosnia, during the UN operations in that area, in which of course the French had a sizeable military contingent.

A successor system, Helios 2 is currently planned, with a projected launch date of 2000. It is intended that the second-generation satellites will be improved, infra-red observation satellites which will offer day/night all-weather observation capability, thus for the first time giving Europe a 24 hour a day, all weather satellite reconnaissance capability. Participation in the Helios 2 programme, is however, still very much under discussion, by a number of countries, a discussion which is explored further, later in this chapter.

Cerise

The final operational military satellite, which is currently at the disposal of a European state, is Cerise. This is a French experimental micro-satellite which uses a wide-band (1-20GHz) hyper-frequency receiver to listen in to radio-electric transmissions and thus allow a database of the radio-electric environment between 600 and 800km altitude to be built up. Costing around FFr 80 million, it weighs less than 50kg and will have a maximum life of around 2.5 years. The data which is collected by Cerise will be used in the development of later ELINT satellites, such as Zenon. Cerise was launched in mid-1995, as a secondary payload on the Ariane launch carrying the first Helios satellite.

VI. Proposed and Future Developments

In addition to the systems outlined above, which are actually in operation, there are a number of other systems, both national and international which are in the planning or development phase. These fall within three broad headings: communications satellites; observation satellites; and electronic intelligence satellites.

Communications Satellites

Sicral

The Sicral system is an Italian satellite communications system, designed to cater for national defence and civil protection applications. Amongst its functions, it will provide strategic and tactical communication networks through a number of fixed, mobile and transportable stations, with a capability for domestic national, European and international coverage for the Italian armed forces. In addition, it will provide an emergency domestic communications capability for the Ministry of Internal Affairs and Civil Protection, and will also be utilised to provide communications support for anti-terrorist and anti-drug activities.

The overall system will be managed by the control centres of the armed forces according to their operational needs, under the supervision of the Defence Joint Chiefs of Staff. Selenio Spazio is the prime contractor for the design, development and construction of the complete space-ground system.

A substantial proportion of the ground station and transponder development for the programme follows on from the technologies developed for the civilian Italsat satellite. It will incorporate anti-jamming technology, and will have a high traffic capability designed to offer reliable, safe and secure military communications over a range of geographical coverage areas, dependent on operational requirements. It is also being designed to be interoperable with the NATO system.[201]

In October 1996. it was announced by Arianespace that a contract had been signed for the first SICRAL satellite to be launched by an Ariane 5 launcher in the latter part of 1999.[202]

Collaborative Communication Satellite Projects

The cost of providing wholly national military communication satellite systems will inevitably continue to rise, and one solution is therefore to establish a cooperative system based on an integrated common requirement. Over the past few years there have been tentative moves between France and the United Kingdom towards increased cooperation in the field of military communication satellites. One of the main requirements in trying to establish a cooperative or collaborative programme is a perceived synchronicty in demand from each of the participants. Largely by coincidence, the satellites which comprise the space-based segments of both the French and the British military communications systems will reach the end of their useful lifetime in the first decade of the next century. Consequently, for some time now, representatives of the UK Ministry of Defence and their French counterparts have been actively discussing the possibility of cooperation or collaboration on a joint successor system to Skynet 4 and Syracuse 2. A series of meetings between the two has been studying the prospects of collaboration in this field.

In addition approaches have been made to several of their other European allies, including Germany, the Netherlands, Spain and Italy regarding the possibility of developing a common military communications satellite system, EUMILSATCOM, and these discussions are still proceeding.

At the same time, cooperation between the French, the British, and the Americans has also been proposed. As well as Skynet 4 and Syracuse 2 reaching the end of their useful lives by the turn of the century, the US DSCS 3 system is also expected to become redundant at around the same time. Again, discussions to assess the viability of such a proposal (nominally designated INMILSAT, standing for International Military Satellite system) have been held over the past couple of years to determine whether they should concentrate on working towards a joint system, or go their own, national ways.

After several years of discussion over various permutations of countries joining together to produce a common, cooperative military communications satellite system, it appears that the favoured option is now a partnership between Britain, France and Germany. In September 1996 it was reported that the defence ministries of these three countries had tentatively agreed to build four such satellites at a cost of $2.6 billion. It was reported that the British and French ministries of defence had written to all potential industrial bidders asking for proposals for the system, which has been designated TRIMILSATCOM.[203] No precise technical specifications for the system as yet known, and it is believed that a contract will not be issued until early 1999.

Reconnaissance Satellites

Osiris

This is a French programme costing some FFr 10 billion. Osiris is a high-resolution, synthetic aperture radar, observation satellite. it is scheduled to complement/replace Helios around 2001-2003. It will offer resolution of 3 metres or better and will be located in a 600-800km polar orbit. Based on a Spot 4 platform, the satellite will weigh around 2.5 tonnes, including the 0.5 tonne synthetic aperture radar itself.

Helios 2

As previously mentioned, it is planned that the existing Helios programme should be augmented with a second generation satellite system, Helios 2, which will offer considerably enhanced observational capabilities using a three satellite constellation. However throughout 1995 and 1996 this programme was been the subject of considerable debate and discussion, principally between France and Germany. The pre-development phase of this project was announced by France in 1995, but the next 18 months witnessed a see-saw debate between the French and the Germans over German participation in the project.

In December 1994 Chancellor Helmut Kohl, of Germany, had said that both countries had agreed that the negotiations [on the Helios 2 system] would be completed successfully by March of the following year at the latest. At the same time, it was reported that a French official said that France and Germany would have agreed an equal partnership on the programme by March.[204] It was thought that a German decision to participate in the programme would encourage the Italians to remain in, and also potentially result in an earlier decision by the Spanish to withdraw from the Helios 2 programme being reversed. It was also hoped that other countries might then choose to become involved. In parallel with such discussions on specific military space programmes, there were continuing talks on the possibility of Aerospatiale and Deutsche Aerospace merging their satellite divisions.

However, such enthusiasm was to prove ill-founded. By the end of 1996 little progress had been made and a political stalemate still stalled the potential merger of the satellite divisions of the two companies. At this stage it was also felt that the involvement of the Germans in the Helios 2 programme remained in doubt. A Franco-German summit between Chancellor Kohl and President Chirac in December 1996, which it was hoped would resolve these issues failed to do so. A statement was issued following the summit, re-affirming that Helios 2, and a later satellite named Horus, would be built early in the next decade, but at a subsequent press conference, Chancellor Kohl said that Germany would be unable to finance the work until 1998 at the earliest.[205] If continuation of service beyond Helios 1 is to be maintained, development work on the successor system must commence in 1997. It is therefore likely that, notwithstanding the lack of international agreement, the French will go ahead unilaterally with the programme, at least in the early development stages, in the hope that other partners will come forward in the near future. Indeed this view was confirmed in one press report in May 1997, in which Jean Yves Helmer, the head of French arms procurement, said that France still expected Germany to join Helios 2 by the end of 1997. The report also suggested that French government officials were intimating that they would build Helios 2 with or without German government support.[206]

Electronic Intelligence (ELINT) Satellites

This is a comparatively new area for European nations. It is also an area which is subject to rather greater secrecy than other areas of military space activity, and thus it is more difficult to find details of the systems and their missions in the open literature. The French Cerise experimental satellite has already been referred to and is scheduled to be complemented by a further experimental satellite Clementine, scheduled for launch is 1998, which, it was planned, would in turn be followed by an operational satellite, Zenon. No further information is known about Clementine at this time, other than that it is a second ELINT demonstrator. The three systems, Cerise, Clementine and Zenon, can therefore be regarded as a complete package of two experimental, proving satellites, leading to the eventual deployment of their operational successor, Zenon. However it must be added that the future of this series is in some doubt due to the current status of the Zenon programme itself, with a programme outlay of some FFr 5 billion proving a little hard to justify in the post-Cold War environment. Evidence of this can be found in the fact that Zenon was originally scheduled for launch around 2000-2005, however, in June 1995 it was announced that all funding for this programme had been cancelled by the French Government and therefore operational deployment at this time must be considered unlikely. Perhaps significantly, it was not the programme itself which was cancelled, but funding for it. It is therefore possible that at some future date, funding could be re-instated. In that sense, the programme itself remains nominally active although its ultimate fate is unknown. In effect the programme has suffered a *de facto* cancellation.

Launch Capabilities

In any consideration of current European military space capabilities, attention must also be paid to launch capabilities. However many satellites may be built, of whatever kind, some means has to be found of placing them into their appropriate orbits around the earth.

As far as European countries are concerned, the actual launching of the satellites is a commercial enterprise. No European country has an indigenous military launch programme and the launch of military satellites is simply purchased from an appropriate launch provider, such as Arianespace, or McDonnell Douglas. This situation is born of purely economic reasons, with no European country in possession of a military space programme, or indeed a civilian space programme, of sufficient size to warrant the highly expensive development of an indigenous launch capability. In terms of assured access to launch facilities, the multinational ownership of Arianespace and its highly successful Ariane series of launch vehicles, provides sufficient guarantees of availability. This situation has proved entirely acceptable to all parties concerned so far and is unlikely to change in the foreseeable future.

VII. Military - Civilian Synergy

There may be considerable overlap between military and civilian programmes. This may take the form of linked programmes, with the most notable example of this coming in the form of the French exploitation of the civilian Telecom satellites to carry the military Syracuse payload. It may simply be the case that technology developed for a civilian programme is utilised in a military one. Once again the French show the advantages that can be gained via the exploitation of dual use technology, with Helios 1 drawing heavily on technology developed and tested in the French civilian SPOT programme. Finally, it may be that the military merely utilise civilian facilities, with the already discussed utilisation of commercial launch services an example of this, with yet another coming in the form of the transmission of less sensitive, long-range communications via commercial channels

A number of countries make use of channels on commercial or national civilian satellite networks for military communications. The US Department of Defense leases spare channels on commercial satellites for a considerable proportion of its routine long-range communications traffic. In June 1995 it was revealed that the Department of Defence expects its satellite communication requirements to increase 5-fold between 1995 and 2010, and the Director of the Defense Information Systems Agency was reported to have said the Pentagon will never be able to buy government unique satellites in quantities great enough to fulfil that level of demand.[207]

Similarly, civilian satellite imagery can be used to obtain useful military data. While the resolution of such imagery may not be good enough to satisfy all requirements, it can certainly be of use. The possibility that such information could be damaging to national security interests is borne out by the belief that US Presidential Directive 37, issued by President Carter in June 1978 deliberately limited the resolution of US civilian remote sensing satellites to no better than 10 metres for this very reason. Indeed the French government currently bans the sale or supply of images with a resolution better than 5 metres.

The final example is perhaps the best known. Signals from the US NAVSTAR/GPS navigation satellites are available to commercial users, with anyone able to buy a GPS receiver, off the shelf. However the signals that can be picked up by the commercially available receiver are deliberately degraded and do not offer the same accuracy as the encrypted signals which are received by military equipment (although they are still quite accurate enough for most civilian purposes). Indeed during the Gulf war, due to a shortage of military receivers, many troops went out and bought civilian ones - so many in fact that the encryption was cancelled so that they could receive the most accurate signals.

Thus, it can be seen that there may be considerable overlap between military and civilian space activity. Some however, feel there should be more. Arguments have been put forward that just as ESA provides a more coherent collaborative basis for civilian space endeavours in Europe, so too might a European Military Space Agency. Indeed ESA, bound though it is by its charter provision relating to the solely peaceful use of space, does not regard all military space activity with horror. It has, for example, launched, or is contracted to launch, Helios, Sicral, Skynet and Syracuse.

There have been calls for greater integration of civil and military space activity in Europe - to deliberately increase the synergy which already exists between the two sectors. This topic will be considered in rather more detail in Chapter Seven.

ENDNOTES

[161] De Maria, M., 'Europe in space: Eduardo Amaldi and the inception of ESRO', *ESA HSR-5*, (ESA Publications Division, Noordwijk, 1993), pp.3-4

[162] De Maria, M., Ibid., pp.3-13

[163] Schichtle, C., 'The National Space Program from the Fifties into the Eighties', *National Security Affairs Monograph Series*, 83-6, (National Defense University Press, Washington DC, 1983), pp.52-53

[164] Healey, D., 'The Sputnik and Western Defence', *International Affairs*, Vol. 34(2), April 1958, p.148

[165] White House Press Statement, *Remarks by the President at intervention for the North Atlantic Council Summit*, (White House Press Office, Washington DC, 10 January 1993)

[166] Asmus, R.D., Kugler, R.L., & Larrabee, F.S., 'Building a New NATO', *Foreign Affairs*, 72:4 September-October 1993, pp.28-40

[167] Asmus; R.D., (*et al)*, Ibid., p.32

[168] WEU Assembly Press Release, 1/94, (Assembly of WEU, Paris, 1994) p.1

[169] Taken from written evidence to the House of Lords Select Committee on Science and Technology, UK Ministry of Defence, *HL Paper 41-II, (HMSO, London,* 28 October 1987), p.147

[170] GEO is a specific orbital plane, located over the earth's equator, at an altitude of around 35,800 km. In this orbit, a satellite appears to virtually 'hover' over a fixed point on the earth's surface, permitting the use of a fixed antenna for relaying signals. It also offers a wide area of coverage, signals from such a satellite being receivable over approximately one third of the earth's surface.

[171] Lee, C., *War in Space*, (Sphere Books, London, 1987), p.85

[172] Jasani, B., *Countdown to Space War*, (SIPRI/Taylor and Francis, London, 1984), pp.85-86

[173] Turnill, R., (ed.), *Jane's Spaceflight Directory 1988-89*, (Jane's Publishing Company, London 1989), p.384

[174] Jasani, B., *Countdown to Space War*, (SIPRI/Taylor and Francis, London, 1984), p.132

[175] Swahn, J., 'International Surveillance Satellites - Open Skies for All', *Journal of Peace Research*, Vol.25 (3), September 1988, p.235

[176] .Tables of the necessary ground resolution to identify/analyse targets appear in a number of sources e.g. Florini, A., 'The Opening Skies', *International Security*, Vol. 13:2, Fall 1988, p.98, in which other sources are also cited, and it is noted that their accuracy has been borne out by other analyses.

[177] Florini, A., 'The Opening Skies', *International Security*, Vol. 13:2, Fall 1988, p.96

[178] Burrows, W., *Deep Black*, (Random House, New York, 1986), p.232

[179] Florini, A., 'The Opening Skies', *International Security*, Vol. 13:2, Fall 1988, p.96

[180] When fully operational, it will consist of a constellation of 24 satellites, 21 active, and three spares in orbit. This will allow full accuracy 24 hours a day

[181] *Space News*, Vol. 2, No.15, Apr 29 - May 5, 1991, p.3

[182] Kaminski, P., quoted in *Defence Issues*, Vol.10, No.41, (American Forces Information Service, Washington D.C., 1995)

[183] Turnill, R., (ed.), *Jane's Spaceflight Directory 1988-89*, (Jane's Publishing Company, London 1989), p.405

[184] Turnill, R., (ed.), Ibid., p.401 and p.426

[185] Turnill, R., (ed.), Ibid., p.401

[186] Turnill, R., (ed.), Ibid., p.428

[187] Turnill, R., (ed.), Ibid., p.406

[188] Jasani, B., 'Satellite Programmes', in Jasani, B., and Sakata, T., (eds.), *Satellites for Arms Control and Crisis Monitoring*,(SIPRI/OUP, Oxford, 1987), p.33

[189] Ford, D., *The Button*, (Unwin Paperbacks, London, 1985), p.61

[190] Gumble, B., 'Airforce Upgrading Defences at NORAD', *Defence Electronics*, August 1985, p.98

[191] Dutton, L., De Garis, D., Winterton, R., and Harding, R., 'Military Space', in *Brassey's Air Power: Aircraft, Weapons Systems and Technology Series*, Vol.10, (Brassey's, London, 1990), p.64

[192] Anson, Sir Peter, *The Skynet Telecommunications Programme*, (Marconi Space Systems Ltd., Portsmouth, 1988)

[193] UK Ministry of Defence, *Statement on the Defence Estimates*, (HMSO CM 3223, London, 1996)

[194] UK Ministry of Defence, *Press Release 141/95*, 7 December 1995

[195] UK House of Commons, *Hansard*, Column 543, Written Answer from Ian Taylor to a Parliamentary Question tabled by Menzies Campbell, (HMSO, London, 4 February 1997)

[196] *Space News*, Vol.5, No.48, Dec 19-25, 1994, p.1

[197] Alcatel Espace, *Syracuse Military Satellite Communication System*, (Alcatel Espace Imformation Brochure, Courbevoie; France, May 1989)

[198] *Aerospace World*, May 1992, p.61

[199] Gallego Serra, F., and Luria, R., 'Spanish military space programs', *International Defence Review*, No.4 1990, p.426

[200] UK Ministry of Defence, *Press Release 141/95*, 7 December 1995

[201] Selenio Spazio, *Information Sheet* (undated)

[202] CNES, *Press Release 96/31*, (CNES, Rome, 30 October 1996)

[203] *Space News*, Vol. 7, No.35, Sep 9-15, 1996, p.1

[204] *Space News*, Vol. 5, No.46, Dec 5-11, 1994, p.4

[205] *Space News*, Vol. 7, No.48, Dec 16-22, 1996, p.17

[206] *Space News*, Vol. 8, No.19, May 12-18, 1997, p.2

[207] *Space News*, Vol. 8, No.23, Jun 12-18, 1995, p.3

CHAPTER SIX

THE WEU AND THE POST COLD WAR SECURITY ARCHITECTURE IN EUROPE

I. Introduction

In the preceding chapters we have outlined the historical development of the Western European Union and, following several years in the international wilderness, its re-emergence as one of the key actors on the European security landscape.

Elsewhere in this book the focus has been on the way that the WEU has sought to supplement its predominantly political activities with an operational capability via the introduction and development of organs such as the Satellite Centre. In addition to such moves, considerable operational practice has been gained through involvement in the international response to the crisis in former Yugoslavia, with Operation Sharp Guard the most notable example. In this chapter the emergence of the new security architecture in Europe following the end of the Cold War and the significant role that the WEU plays at the heart of adaptation to this new era is explored. Thus we investigate the inter-relationship of the WEU with both NATO and the European Union, with the emphasis on the desire to deliver some form of European operational capability in the light of shifting American perceptions of national interest. This account will then form the backdrop for the final chapter in which we seek to assess whether the development of these capabilities can be seen to signal a decisive shift in favour of a re-Europeanisation of defence and security, resulting in an eventual move away from reliance on the United States and NATO; or whether, with NATO still representing the main institutional linkage between the states of Western Europe and the United States of America, the transatlantic linkage still has some mileage left in it and this increased level of European dynamism in provision of defence and security functions merely reflects a much needed move towards equalisation of the seurity burden in Europe.

The evidence presented thus far makes a compelling case for the conclusion that the European states through the WEU seek to define for themselves a more influential role in the future conduct of their own security policy. However, this task is far from simple and in the confused institutional climate we find ourselves in, it is widely intimated that the development of such a role forces a prioritisation within the organisational hierarchy.

Thus we can see that the continued existence of the WEU in itself is seen as providing two possible options as regards its ongoing role in European security: either the WEU can be classified as a distinct alternative to the existing transatlantic framework, giving Europe back the lead in its own defence through the eventual realisation of common defence and security policy co-ordination within the European Union (EU); or, alternatively, it can be considered a central pillar of a system of interlocking institutions, with NATO, the EU and WEU operating alongside each other.

It will be the contention of this chapter that the latter of these options is both the most prudent option available and the predominant viewpoint on both sides of the Atlantic.

However, before moving on to look at the way in which the organisations have sought to reposition themselves following the end of the Cold War, it is worth outlining the nature of the radically altered strategic and security environment with which all of these organisations are now faced.

II. The New Security Environment

There is a widespread consensus among commentators and politicians alike that the nature of military operations and the type of missions that national forces will be expected to become involved in is undergoing a major shift in emphasis. Gone are the darkest days of the Cold War when military planners were tasked with the challenging issue of how to deal with the apparently ever present threat of Russian hordes crashing across the German plains and into mainland Europe. Whether or not this was ever a realistic vision of the challenge which NATO was likely to face is now a matter of historical debate. What is clear is the fact that the forces and plans formulated for this contingency, and the challenge that it presented, are no longer valid and there exists a pressing need to reassess how the structures put in place for just such an eventuality can be expected to cope with an entirely new agenda of potential security problems.

This point in itself has provided much fuel for those who wished to toll the death knell of NATO, perceiving it as an alliance that was incapable of adapting to the rigours of its new environment.

> 'No alliance has ever persisted longer than the threat that spawned it. The idea of an alliance is logically inseparable from the idea of a threat and a foe ... In the absence of a threat and a foe, the NATO alliance as we know it will not persist much longer.' [208]

This was a commonly held view in the early 1990s at the time the process of adaptation was beginning. Yet today NATO is still very much alive and kicking, firing its first shots in anger during the United Nations mission in Bosnia, despite having traversed 40 years of Cold War without ever doing so. Indeed in the 1990s NATO is now looking to become more active than it ever has been. Despite this, there still exists an attitude that it is the wrong tool for the job of European security in the next century. Strangely, although the WEU shares a similar lineage to NATO, it is not generally seen to share this inability to adapt. Whereas many would argue that NATO is a relic of history, the WEU, aided by its association with the EU, is looked on as a different breed altogether. With the constitutional advantage that it requires no 'external threat or adversary to justify itself' and as the potential defence component of the EU, it is situated to become involved in the European Union's inevitable input into areas which have military implications and thus the WEU could hardly be better placed to take advantage of the upheaval in European security.[209]

That a new set of security institutions is called for in certain quarters is predicated upon the inability of the old order to deal with the challenges which will be presented in the future. If the immediate post Cold War period is anything to go by, the most notable type of missions will be those directed towards the increasingly common arena of internal conflict. While this is an area which the international community has found outrageously difficult to deal with, there is growing pressure on international organisations to become involved in these crises in an attempt to tackle both the causes and consequences of this type of conflict.

In the past this type of situation proved problematic because of the ideological differences which made any involvement in the domestic situations of states politically provocative and rendered such areas a minefield which IGO's entered at their considerable peril. The zero-sum politics of Cold War ideological rivalry meant that if one side were supported by the Americans, the other would almost invariably receive some support from the Soviets. Luckily these proxy conflicts are a demon of the past and there is now enough consensus to fuel the feeling that something can and must be done to relieve the suffering of populations under threat in civil and ethnic conflicts.

As a result of this sea-change in the relationship between the former Cold War enemies, we are likely to find the type of missions carried out in Bosnia jointly by the United Nations, NATO and the WEU and supported by the Russians, as more indicative of the new norm of military activity, with humanitarian intervention, peace-keeping and in certain instances peace-enforcement likely to supersede notions of war fighting. Former United Nations Secretary General, Boutros Boutros-Ghali in his 1992 Agenda for Peace,[210] helped define this new agenda when he spoke of the several tasks which faced the international community in terms of what had previously fallen under the broad heading of peace-keeping. This was of additional relevance to the organisations of Western Europe because of the emphasis which the 'Agenda' places upon subcontracting missions to regional organisations as a crucial element in the international community's ability to come to terms with the strategic situation of the late twentieth century and beyond:

> 'Consultations between the United Nations and regional arrangements or agencies could do much to build international consensus on the nature of a problem and the measures required to address it. Regional organizations participating in complementary efforts with the United Nations in joint undertakings would encourage States outside the region to act supportively. And should the Security Council choose specifically to authorize a regional arrangement or organization to take the lead in addressing a crisis within its region, it could serve to lend the weight of the United Nations to the validity of the regional effort. Carried forward in the spirit of the Charter, and as envisioned in Chapter VIII, the approach outlined here could strengthen a general sense that democratization is being encouraged at all levels in the task of maintaining international peace and security, it being essential to continue to recognize that the primary responsibility will continue to reside in the Security Council.' [211]

With this desire to enhance the provisions of Chapter VIII of the UN Charter, which specifically outlines the role that regional organisations can play within the United Nations framework, at the very heart of Boutros-Ghali's vision of how the UN can cope with the flood of peace-keeping tasks which it finds itself forced to deal with, it is unsurprising that these tasks have also come to figure prominently in the defence planning of the member states.

This new security environment and the changed emphasis in the type of mission likely to be conducted, has resulted in considerable alteration in the rhetoric used to justify military expenditure and to the way in which the states prioritise military activities. The French for example have sought to highlight their commitment to post Cold War tasks by reformulating the way in which their defence budget categorises various activities, including allocations for roles such as peacekeeping and space, a change from the rather less scrupulously justified budgetary considerations of the Cold War.

That states are acknowledging these differences through reduced troop numbers, more mobile forces, retraining and altered battle plans is reflected in the activities of the organisations to whom these troops are ultimately responsible. The very fact that NATO was eventually able to put a force into Bosnia, contradicting the notion built into the Washington Treaty that NATO was an organ of collective defence and thereby effectively ruled itself out of the type of out of area activity which Bosnia represented, demonstrates a willingness on behalf of the membership and the organisation to move forward and embrace the new tasks being set down for it and is proof, if proof be needed, that times have indeed changed.

Part of this desire to adapt to the changing strategic environment can be seen in the efforts on behalf of the European's to take more control in situations which require the provision of purely European operations. As suggested above, NATO (and with it the Americans) did eventually become involved in the peace-keeping activities in former Yugoslavia, something which at the time appeared to signal a failure on behalf of Europe to assert itself in the first major test of its resolve. However, NATO involvement need not be viewed as a European failure, rather it should be deemed a valuable learning experience. Following the attempts of the European Union and the OSCE to find a more predominantly European response to the crisis, the NATO involvement proves several things, most notably that NATO can adapt to a different type of operation than that for which it was designed, as well as pointing up the deficiencies in the current European approach to security and defence co-operation. This latter is something which will only be rectified through time and in conjunction with NATO. As has been widely argued, even if some kind of European consensus could have been achieved over what to do in the Yugoslav crisis, it is very unlikely that the operational capacity or organisational framework existed within which the Europeans could have conducted such a mission, with the principal candidate, the WEU, at that time far from operationally credible.

Despite the technical and logistic barriers which currently hamper the aspirations of the WEU and which Bosnia highlighted, the organisation has accepted the logic of the post Cold War environment and acknowledged the type of tasks it may have to become involved in:

> 'Apart from contributing to the common defence in accordance with Article 5 of the Washington Treaty and Article V of the modified Brussels Treaty respectively, military units of WEU member States, acting under the authority of WEU, could be employed for:
>
> - humanitarian and rescue tasks;
> - peacekeeping tasks;
> - tasks of combat forces in crisis management, including peacemaking.' [212]

In addition to these tasks which may demand the use of an active military intervention, there are several other types of security risk which the WEU points to as areas of concern. These are primarily of a transnational nature and are generally viewed as problems which can only be effectively dealt with on a multi-lateral level. These include the potential for 'proliferation of weapons of mass destruction and their means of delivery', the threats posed by 'international terrorism, organized crime, drug trafficking and uncontrolled and illegal immigration' and 'environmental risks'[213]. Compared to the type of peace-keeping mission discussed earlier, these alternative considerations on the security agenda are by no means entirely new, yet they present a very different set of challenges. Some demand political solutions, whereas others have a clear economic element and there may indeed be some which necessitate a form of military response. All of them command attention, yet it remains the humanitarian and peace-keeping missions which thus far have captured the attention of the media and which form the core of what NATO and WEU are likely to become involved in operationally.

In the interim period it is unlikely that such missions will be deployed under a purely WEU mandate, with the most likely outcome being some kind of *ad hoc* co-operative action with NATO, at least until the WEU is confident and capable enough to go it alone. Ultimately it is fully expected that the

WEU will have the capacity to launch operations which are deemed to be of purely European interest and this is entirely in keeping with the emerging transatlantic relationship which will be discussed below.

In recent years there has been a flurry of diplomatic activity in an attempt to bring the organisational structure into line with the altered security situation of the 1990s. This has included much discussion over the development of a European approach to security issues; either through the development of European Security and Defence Identity (ESDI), seen as a means of strengthening the European wing of NATO; or via the development of a Common Foreign and Security Policy (CFSP) which is seen as the next stage in the development of the European Union; and including attempts to provide a vehicle for European operational capability via the Combined Joint Task Forces (CJTF) concept, which should eventually allow just the kind of European (most likely to be WEU based) operation envisaged above. All of these can be seen as providing an alternative to, but not a replacement for, NATO and the transatlantic linkage which it provides.

In the following section we will trace the development of the new security architecture in Europe and the consolidation of organisational diversity which it has resulted in.

III. The Emerging Architecture

The traditionally accepted security situation in Western Europe is undoubtedly changing. The decision at Rome to re-engage the exclusively European WEU may have been informed by the strategic circumstances of the time and disillusionment with the exclusion which European states felt from the superpower dialogue, but it can also now be seen as a decision which has allowed the Europeans to develop an understanding which was perhaps always more suited to a post Cold War security environment than it was to the rigidly defined bloc pattern of the mid 1980s. The WEU after all has been allowed to flourish as never before and now stands alongside the two major organisations of Cold War Europe.

Yet there is evidence in the Rome Declaration itself and subsequent documentation which belies the progress that the WEU has made in its attempt to stake a claim for influence in the security structures of Europe. In 1985 the WEU Assembly through its system of Rapporteurs, commented that the WEU:

> '... is not and should not become a military organisation. The only possibility at present is to integrate European forces in NATO or to maintain forces under national command...and it is still impossible to envisage a defence union, in the context of the European communities, for instance ... WEU is destined to remain a political organisation... it is therefore by political means that WEU can help build the European pillar of the alliance, or even form a pillar of the alliance itself ... '[214]

This accurately reflected the feeling of the time that NATO should remain the only organ responsible for the application of West European security policy. However, following the upheavals of the late 1980s, this attitude began to change and the desire for some level of re-Europeanisation in the realm of defence took hold. Subsequent decisions have allowed this vision to move forward dramatically. Immediately prior to the Maastricht Summit of December 1991, the Heads of State and Government meeting of the North Atlantic Council met in Rome, where, among other things, they affirmed the need for a more coherent European role in NATO as part of the Alliance's attempts to adapt to the changing strategic environment:

> 'The development of a European security identity and defence role, reflected in the further strengthening of the European pillar within the Alliance, will reinforce the integrity and effectiveness of the Atlantic Alliance. The enhancement of the role and responsibility of the European members is an important basis for the transformation of the Alliance.'[215]

The ESDI was further promoted by the subsequent Maastricht Treaty, which highlighted the need for the Europeans to 'develop a genuine European security and defence identity and a greater European responsibility on defence matters'[216] and is something which is now widely accepted within the Alliance where:

> 'NATO leaders agreed that the emergence of a European Security and Defence Identity (ESDI) would strengthen the European pillar of the Alliance while reinforcing the transatlantic link. They authorised the further adaptation of the Alliance's political and military structures to reflect its new roles and missions and the development of the emerging ESDI, including the development of separable but not separate capabilities that could be used by NATO or WEU.' [217]

Despite assertions of this kind coming at the highest Ministerial level, the ESDI remains an ethereal concept. To some it means much more than merely the attempt by the Europeans to become a cohesive unit within the Alliance structure. For this group the ESDI can be seen as the first step towards a fully integrated and independent European defence, perhaps based around the WEU, perhaps as part of the EU. However, despite the ongoing French negotiations over their re-entry into the integrated military structure of NATO providing some scope for those who consider French reluctance to participate in the full activities of NATO as some beacon to light the passions of those who would see Europe cater for its own security,[218] this particular vision of the ESDI may be waxing, with the more commonly perceived view of ESDI as a means of bolstering, not undermining, the Atlantic alliance, asserting itself as predominant.

IV. Maastricht and the CFSP

The entry into the European psyche of the concept of an ESDI, was accompanied at the same time by an attempt to formalise the WEU's role in the development of the emerging security architecture. This process was facilitated through the inclusion of passages in the Maastricht Treaty which sought to chart a course for the direction of relations between the WEU, the EU and NATO.

In an effort to move the integration of Europe forward, Maastricht broadened the scope of the 1957 Treaty of Rome and under the guise of the renamed European Union, added to the predominantly economic vision that had gone before. No longer was the European adventure to be limited to such mundanities as agriculture and fisheries policy, Maastricht not only advanced the prospect of an Economic and Monetary Union, it supplemented the existing European Communities superstructure with two additional pillars: one being the attempt to harmonise Justice and Home Affairs; the other the introduction of the notion of a Common Foreign and Security Policy (CFSP).

These two pillars were and remain purely inter-governmental, with no power being transferred from national governments to Brussels in these most sensitive of areas. While the grand federalist vision of a united Europe outlines cooperation in such areas as the very core of deeper integration, no commitment was being implied on behalf of the member states at this early stage. Indeed, even several years down the line, there is no obvious prospect of policy in these areas joining the realms of issues which can be decided by anything other than unanimity among the members. However, the concept of a CFSP is widely perceived as central to any future development of political cooperation within the Union, underpinning as it does the security of the member states and thereby projecting a common outlook on a matter of paramount importance to all participants. Without the ability to co-ordinate activities in this area, it is unlikely that any further deepening of the political linkages between member states will be possible. Thus the Treaty highlighted the importance of CFSP and the desirability of a developing European Security and Defence Identity (ESDI), within which the gradual realisation of common aims and goals could be achieved.

As part of this, a crucial role was identified for the WEU, essentially serving as a bridge between the military activities of NATO and the political activities of the EU.[219] Having thus, openly acknowledged the need for the WEU to move beyond the purely political role it had fulfilled for the bulk of its existence, the Maastricht Treaty paved the way for both the organisation's further development and the widespread acceptance of the organisational diversity which now characterises European security.

The Combined Joint Task Force Concept

With the dual efforts of ESDI and CFSP now in the public arena and with the re-activated WEU serving as a vehicle for the eventual realisation of both, there remained an additional consideration which had still to be tackled - the perceived inability of the West Europeans to co-ordinate and execute a mission of their own volition. While progress towards an ESDI provides the institutional basis within NATO for some form of European cooperation and the CFSP is clearly intended to provide the political will behind such a concept, as we saw above, the WEU had ruled out the possibility of participating in military missions as recently as 1985 and when called upon to participate in the Yugoslav crisis, was singularly incapable of doing so in anything other than a supporting fashion.

The desire to avoid any charge of duplicating the activities of NATO had prompted the WEU to carefully avoid the politically indelicate matter of its own operational viability. Fearful of being seen to overstep the mark and perhaps also afraid of showing too much ambition in this area, the WEU in the first phase following its reactivation essentially served the political purpose that is now being accorded the concept of an ESDI, namely that much be made of the necessity that the WEU serve as an obedient European subset within the wider Alliance. Thus the WEU states had proclaimed that:

> 'We intend therefore to develop a more cohesive European defence identity which will translate more effectively into practice the obligations of solidarity to which we are committed through the modified Brussels and North Atlantic Treaties.' [220]

Although this refers to both Treaties and both organisations, it is clear that the retrenchment of European commitment to NATO at this stage gave the broader body the central role. With the onset of the post Cold War era this attitude has been challenged and some form of military independence is deemed vital. As a consequence of this change, and existing alongside the political moves made in both Rome and Maastricht, one of the most crucial areas in the development of the WEU and its ability to provide a coherent and credible response to the type of challenges outlined above, has been the gradual move towards acceptance of the Combined Joint Task Forces (CJTF) concept.

As part of the mechanism through which the states of Europe and North America have sought to adapt to the changed strategic environment, the CJTF provides a versatile vision of the way forward in the provision of security operations. With the new security situation necessitating a different pattern of forces to combat new generations of problems which may arise, while at the same time maintaining the capacity to defend the member states of the Alliance from the resurgence of a more traditional threat, it was apparent that something much more flexible was required than the strategic doctrines of the Cold War.

The resultant solution, the CJTF, is neatly summarised as a:

> ' ... deployable multinational, multi-service formation generated and tailored for specific contingency operations. It could cover a wide range of potential tasks including humanitarian relief, pecekeeping or peace enforcement. The employment of a CJTF for Article 5 operations is also not excluded.' [221]

Thus we are looking at a highly mobile, diverse, well equipped and truly international force, capable of being adapted to the individual circumstances of the current situation under consideration. It envisages a role for not only the existing NATO members, including the Americans and the Canadians, but also suggests that, given the correct set of political and security considerations, an operation based around only the European members could be launched, ostensibly led by the WEU.

Having been formally endorsed at the Alliance's Brussels summit in January 1994, the concept has had some time to bed down and is now recognised as central to the development of both an operational capability for the WEU and the broader vision of a European pillar within the Alliance.

Although the CJTF is primarily envisaged as NATO's response to the security challenges of the next century, circumstances may become apparent where additional involvement is sought from outwith the NATO area, or alternatively a situation may arise where the United states is unwilling to participate. Thus it is entirely plausible that a Task Force could be drawn entirely from within the European membership of NATO and placed under the control of the WEU. In addition to this, it is envisaged as the most likely mechanism through which NATO can utilise assets from states outwith the Alliance structure as and when these are required. All of this outlines a flexibility which is particularly useful when one considers the type of operation that is now being envisaged, with the need to integrate troops from other states vital if truly international humanitarian and peacekeeping tasks are to be performed.

As an example of the kind of thing which can be developed further in the future we can consider the operations performed by NATO in the former Yugoslavia following the Dayton Peace Accords. Although not a CJTF operation by name, the Implementation Force (IFOR) which NATO sent to Bosnia and its successor the Stabilisation Force (SFOR) are both possible examples of what the broader vision of CJTF hopes to achieve. The inclusion of forces from sixteen[222] nations outwith the existing membership of NATO, including Partnership for Peace and non-European states, has provided much needed operational and co-operational lessons from which NATO can build for future combined actions.

Of course the situation in Bosnia was by no means an example of successful policy making among the European allies. Reflecting on the situation prior to the despatch of troops to the region, the then NATO Secretary General Willy Claes remarked:

> 'It was indisputable that Europe did not shine with its common vision at the beginning of the tragedy (in Bosnia). But it is no less true that for a period of time the United States retained a certain distance from the problem. In this phase ... we cannot deny their primordial role. Europeans should not take this as an affront but as a stimulation to set up their own common foreign and defence policy.'[223]

However, it was the eventual involvement in Bosnia which best highlighted the type of action that may be possible under the aegis of CJTF. The adaptability and potential universality of operations conducted in this way are seen as being the most beneficial elements of the CJTF, allowing the North Atlantic Alliance to deal with the rigours of the current era whilst also laying claim to being a truly international response to the interdependent security considerations of the age.

Very much part of this grand scheme is the aforementioned inclusion of former Warsaw Pact countries in the Bosnian forces. This involvement of former enemies represents the next area we will consider in the emerging security pattern in Europe, with the moves towards enlargement of all the organisations gathering pace proving that progress will not be hampered by the intransigence of the Russians. We are rapidly moving to a situation where all the states of Europe are involved in the defence of the continent.

V. A Pan European Security Architecture

One of the most important features of the post Cold War security environment in Europe, is the breadth of the commitment to inclusion of the countries of the former Soviet bloc into organisations which have traditionally been bastions of Western privilege and economic superiority. The most obvious examples of the desire to bring these states into the fold of responsible nations come with the evidence of the dual enlargement debates currently being conducted by NATO and the EU. While these processes are unlikely to be mirrors of each other - the EU for example appears unlikely to make realistic moves towards enlargement before EMU is in place, something which in itself will struggle to meet the deadlines of the twentieth century - it has become increasingly clear that the process of enlargement is inevitable and desirable. The desire to allow former Warsaw Pact countries to accede to NATO is now broadly accepted, with the notable exception of Russia, and the likelihood is that NATO's Madrid Summit in July 1997 is likely to, in Secretary General Solana's words, 'mark the moment when the process of transformation and change converge and the new Alliance finally emerges' he stage was unable to contemplate the extension of Article 5 security guarantees, partly because of the negative image this would have portrayed in the minds of the Russian leadership but also because it was felt that the potential members were economically, politically and militarily incapable of fulfilling the responsibilities of full membership. The PfP was seen as a staging post, providing an interim measure as the states struggled to cope with their individual experiences of transition. All of these circumstances demanded incremental change rather than radical shifts in the European power balance, and the PfP can be viewed as a worthwhile exercise in contingency planning, attempting to keep everyone happy by giving none of them what they wanted but allowing them all to believe they had had a measure of success. Thus the East Europeans were encouraged to become meaningful partners, the pro-enlargement camp could see that their preferred option had been delayed but not forgotten and the Russians were at least not accorded the humbling political reverse that would have been the result of a swift process of inclusion.

For the East Europeans this was neither the membership which they sought, nor was it a brush off, disavowing any current or future likelihood of acceptance into the organisation. However, it was not only in the capitals of the East European states that the PfP was seen as a second class option. The announcement of the Partnership for Peace was greeted with some scepticism at the time, seen as a meaningless construct which would merely duplicate the pan-European organisational membership of the Organisation for Security and Cooperation in Europe (OSCE). Yet despite such criticism, those for whom it was developed, the East Europeans, have made it their mission to be accepted within the organisational framework and to make the PfP work positively for them. Accepting the minimalistic role within the organisation which the Partnership allowed, their already keen involvement in the North Atlantic Cooperation Council (NACC) and participation in the Bosnian operation, have helped prove to the NATO membership that the the accession of certain states need not be feared. The period of acclimatisation and accommodation which followed on from the NATO Summit of 1994 has allowed a deeper understanding between the partner countries and NATO to develop and has brought us to the brink of welcoming the first new full members into the organisation since the accession of Spain in 1982.

Although it has taken several years for the process to mature, it can be argued that the PfP and its alternative to full membership provided a de facto security guarantee for these states which belied their real status within the organisation. The advantages of this variable organisational geometry, with different levels of membership for different states, has been wholeheartedly endorsed by the WEU and is by no means unique to the Atlantic Alliance. The WEU has also set out to acknowledge the security concerns of the East European states by expanding its membership to include these states as 'Associate Partner' countries. With the ten full members and the eight members who participate as either Associates or Observers, the inclusion of nine Partner states took the organisation to a total of twenty-seven participating states by 1996.[226]

This process was formalised at the Kirchberg Summit in May 1994 where it was asserted that:

> '... the development of closer relations with the nine Central European nations through an enhanced status within WEU and through co-operation within the Alliance framework, in particular the Partnership for Peace programme, will be mutually reinforcing and contribute significantly to security and stability in Europe.'[227]

The first involvement of the Associate Partners came at the Noordwijk summit in November 1994, where the involvement of the Eastern states was welcomed:

> 'Ministers feel that the participation of Associate Partners in the meetings of the Council, according to the Kirchberg arrangements, contributes to a true strengthening of WEU and they are convinced that their greater participation in WEU activities and closer consultations and further co-operation on security questions are instrumental in enhancing stability in Europe.'[228]

This inclusion of the East European states in the structures of all the major organisations, has provided an extended period of 'engagement', with the suitors given a chance to assess their compatability as alliance bedfellows. While the formal status of the former Warsaw Pact countries remains limited by the dictates of second class membership, they have proved valuable partners and willing participants in activities ranging from military exercises and standardisation as part of the PfP framework, through to discussion of issues such as defence conversion and air traffic management which fall under the remit of the NACC.[229] That their ability to participate in the discussion of crucial issues regarding defence and security is limited in political terms and although the fact remains that there is no security guarantee *per se*, the progress towards meaningful accommodation has been steady.

Overall, the involvement of the Partner countries in the WEU (the Partners can participate in the Council of Ministers where they are accorded the right to participate but cannot block a decision which is accepted by the ten full members)[230] and a broadening of the scope of NATO activities has provided the basis for optimism as regards the development of all these organisations.

Yet despite such optimism, the main stumbling block to NATO enlargement was and remains Russian fears of a system of Western security Alliances on its doorstep, an eventuality which it deems too close to areas which have traditionally been considered part of the Russian security sphere. For these reasons expansion of NATO has been a topic which the planners have carefully sought to involve the Russians in, extending countless invitations to join PfP and from there to discuss options deepening the involvement of Russia. That Russia is still jittery about the proposed expansion is evidence of the precarious political situation in the former Soviet Union, although it does seem that finally the West is losing patience with the procrastination of the Russians and will force the issue forward in Madrid regardless of potential Russian protestations. Whether this will force the Russians to become more closely involved in the actual negotiations via some form of 16 + 1 talks will depend on the resolve with which they hold reservations regarding expansion. That said, there is now a level of inevitability about the future shape of European security and it is based around a set of bigger, broader organisations. With both the EU and NATO here to stay, and an increasingly entrenched position for the WEU, there would appear to be little chance of the Russians delaying the process any further. The future security architecture of Europe is therefore likely to be very different to the collective defence structure which existed before.

VI. The Status of the United States-Europe Relationship

After nearly half a century of reliance on the security guarantees of the United States, many feel that Europe can and should assume responsibility for its own defence. As we have outlined earlier, the machinations at Maastricht and the moves to allow the WEU an operational capacity, albeit within the framework of pre-existing NATO structures, does provide evidence of, if not a desperate keenness for, certainly a tendency to accept, the logic of Europeanised defence. The development of an independent data interpretation capacity clearly outlines the direction that is being taken. The heart of the CJTF idea is for Europe to accept ownership of its own security and defence interests where possible. If we are to accept that the type of operations which states will become involved in as we enter a radically different strategic environment are significantly different from those which the founders of NATO envisaged, then there is also scope to accept the proposition that perhaps NATO is not the organisation best placed to accept the challenge of dealing with these.

The relative success of the 'Partnership' concept and the acceptance of organisational diversity which characterises the current situation make for a level of optimism about the future of NATO. However, this is a situation which is not without its detractors. Chief among the dissenting voices is a small, but increasingly vocal section of the United States foreign policy elite which points to the ongoing American involvement in European security as an outdated and unnecessary drain on precious resources.

This same caucus of opinion would have us believe that the likelihood of prolonged United States activity in Europe is slim in the extreme, and that the present situation cannot be defended by the American political elite in the long term. They point to the current level of American involvement in NATO as damning evidence of a fundamental lack of foreign policy direction following the end of the Cold War.[231]

This image of a United States ready to leave Europe to its own devices has developed a moderately high degree of political and academic momentum, despite running counter to the stated policy of the administration. With a more isolationist foreign policy for the United States viewed by many on the political right in the United States as a necessary prerequisite for the preservation of economic success which is currently being challenged by the burgeoning economies of the Far East, it is a theme which is unlikely to disappear.

For many it is imperative that the USA not allow itself to be sucked into a global security role which it is felt should be performed by more than just the world's only remaining superpower, and if this advice is ignored it can only have biting economic consequences as yet again the Americans help foot the bill for European security.

On a less anti-NATO note but still critical of the chosen policy direction, is the opposition to expansion which is emerging from the United States. Citing the unknown political and financial costs of expansion, including a near inevitable diminution in the American public's attitude towards Europe as the bill for integrating new members becomes known and the Russians willingness to accept the post Cold War settlement wanes with this first imposed reverse, it is argued that enlargement is a lose-lose scenario for America.[232]

At the heart of this vision is the readiness to ignore one of the fundamental tenets of United States foreign policy in recent years, the desire to see a Europe that is secure and which can serve as both a military and an economic ally. In the past the military guarantees contained in Article V of the Washington Treaty have been the most public manifestation of the United States involvement in the conduct of European affairs. Yet this belies the depth of the alliance, something which on the outside is designed to provide for the physical security of the continent but can be seen as providing security on several levels, including political and economic.

As has been discussed previously, the Americans proved reluctant to become involved in the provision of military security for Western Europe, and their primary concern following World War II was to provide for the rebuilding of war ravaged economies and to get Europe back into shape. This economic agenda was designed to aid security by helping to guard against a descent into chaos and by putting the liberal economic agenda back on course provide benefits for all sides. The resultant Marshall Plan became the first example of what was to become a permanent American involvement in the development of post Cold War Europe, and it was only later when the need to provide a physical security umbrella became obvious, that the military element of the transatlantic link was put in place.

There is no denying that the United States wishes to scale down the involvement which it has in European security. A cursory glance at the troop levels in Western Europe since 1989 shows that the United States is unwilling to maintain the Cold War level of involvement. This is merely prudent and while in the immediate aftermath of the Cold War much was made of the speed of the American withdrawal and discussion focused upon how much of a presence would constitute a 'firebreak' in the event of hostilities breaking out, this logic has now been surpassed by the need for a sensible reappraisal of what the United States commitment to Europe means.

What appears certain is that the North Atlantic Alliance is here to stay and that the United States is both willing to remain involved in European security and is most welcome as an ally. This is a view reflected in positions of influence on both sides of the Atlantic. Dr. Klaus Kinkel, the German Minister of Foreign Affairs commented in 1996 that the preservation of the transatlantic linkage was crucial, not only because of the security imperative but also because of the deep and binding ties of economics, commerce, culture and science.[233] Further to this Dr. Kinkel added that the interlinked security network that we are now faced with does not detract from the role that NATO plays, rather he suggested that the transatlantic nature of the organisation is even more important in this phase of European security, serving as it does as an 'anchor of stability'.[234]

In similar vein, 1996 also saw the United States Secretary of State, Warren Christopher, in commemoration of the 'Speech of Hope' delivered by his predecessor James Byrnes in 1946, talk of his vision for the twenty first century, one which had NATO (and by definition the United States) at its heart as the security pillar. He also reiterated the first principle laid down by Byrnes all those years before - that the United States would never again withdraw from Europe and would remain a political and military power in the region.[235] Although such sentiments do not represent the views of all those in the US, the re-election of President Clinton should provide hope for those who wish to see such sentiments remain at the heart of US policy towards Europe. In this extract from his 1997 'State of the Union' address, President Clinton emphasises the linkage between the security of the continent of Europe and the situation which the United States finds itself in:

> '... we stand at another moment of change and choice - and another time to be farsighted - to bring America 50 more years of security and prosperity. In this endeavour, our first task is to help build , for the first time, an undivided, democratic Europe. When Europe is stable, prosperous, and at peace, America is more secure.' [236]

Given this iron linkage between American and European fortunes, it is not unusual to find that Clinton went on to call for the expansion of NATO by 1999 and the strengthening of the Partnership for Peace, as well as the need to conduct a meaningful dialogue with Russia in the hope that a lasting and peaceful relationship between NATO and a democratic Russia can be built. This belief in the preservation of the transatlantic linkage (and the North Atlantic Treaty in particular) is evidence of the desire to maintain what the United States sees as the primary element in the security of Europe since 1949. However, it makes no mention of the role that the WEU is expected to play within the European security structure, although this too has been the subject of praise at the highest levels within the United States administration. In 1993, Walter Slocombe, then principal under secretary of defense for policy, remarked that the US welcomed the 'emergence of the WEU' and went on to point out that the WEU is regarded by the US 'as a complement, not a rival, to NATO' and is expected 'to develop in close symmetry with NATO's own development'.[237]

The evidence of this is clear, with the United States keen to promote the intertwined notions of CFSP, ESDI and CJTF, all of which will see the European group of states capable of launching operations without the explicit involvement of their North American allies.[238]

What becomes clear from this look at the development of the post Cold War security architecture in Europe is that at the heart of the situation there lies a simple logic: the Europeans want to garner more control of the military aspects of their involvement in alliance activities, but wish to do so while keeping the United States involved in Europe and happy with the role they play. Essentially this means that nothing can be done which would prompt American disengagement from Europe. The United States on the other hand wishes to allow Europe to take responsibility for certain tasks and to lessen the burden of extraneous overseas defence spending it bears, although it is also keen to maintain a role in the defence of Europe and the North Atlantic. These two visions are by no means contradictory, in fact they mesh together quite comfortably and form the natural backdrop to the multi-layered, multi-organisational approach to European security we can see developing.

However, in certain quarters the complementary nature of these visions is not fully appreciated. Despite widespread support for the emerging pattern, there persists a vision of 'Europeanisation' of defence, at the expense of the American involvement, as an issue which has achieved maturity and which could provide scope for the eventual failure of the alliance. Yet given the centrality of the ESDI in the future which has been mapped out for NATO, and the widespread desire to empower Europe to take control of certain areas through the provision of a CFSP and the utilisation of CJTFs, it looks very much like the 'Europeanisation' of defence is an issue which NATO has adopted as its own.

It does seem counter intuitive to argue that the time and effort which has gone into building the interlocking institutions model should be wasted and some purely European replacement be conjured up to replace it. As long as both sides perceive gains from the relationship, those who find the preservation of the transatlantic linkage inimical to the development of a more robust European approach to its own defence are likely to remain in the minority.

The rationale for this opposition to the developing security architecture is rather mystifying, with the logic supporting preservation of the current system of interlocking institutions all but undeniable. Coupled to this there is widespread support for the maintenance of the transatlantic linkage. As we have already discovered, there is no discernible desire among the political leaderships of the member states to see an organisation which has helped secure the peace of the continent disbanded. Indeed the reverse of this is true, with most accepting the logic of both NATO survival and United States involvement as givens in the strategic environment of the 1990s.

None of the principal players want Europe left alone to take care of all security eventualities. What they do want is a Europe which is responsible and mature enough to operate independently of its American allies if or when required. The recent decision by the French government to negotiate a renewal of their linkages with the military aspects of the North Atlantic Alliance has been informed by this very desire. The French, the most ideologically committed to a Europe capable of taking care of its own defence interests, make an unlikely advocate of an alliance traditionally dominated by the link to the USA. However, with the increasing likelihood of co-existing organisations taking responsibility for different types of operation, the French have realised that if they are to have an active input into the future shape of operations they have to be involved in the military activities as well as the political. Charles Millon, French Minister of Defence talks of the need to embrace the political concept of a 'European pillar' and is vehement in his conviction that this is not a challenge to the survival of the North Atlantic Alliance.

> 'There can be no European defence policy that does not take account of the Atlantic dimension; there can be no lasting Alliance without the affirmation of a strong European pillar. The purpose is to arrive at a new form of collaboration between Europe and North America within NATO.' [239]

This is entirely consistent with the notion that Europe can and must take some responsibility for its own defence while maintaining a linkage with historical, cultural and economic allies across the Atlantic.

VII. Conclusions

This examination of the WEU and the position it finds itself in as merely one of a plethora of organisations providing some form of security function within the context of Europe, has helped to outline the complexity of the current security environment as well as raising the question that perhaps the organisational framework within which all of these bodies operate, could be significantly trimmed in light of changes in the threats facing late twentieth century Europe.

While the Maastricht Treaty helped define the position of the WEU as an organisation and placed the ESDI at the heart of the European approach to defence matters, it was the decision to develop a European, Common Foreign and Security Policy (CFSP) which attempted to broaden the horizons of European political cooperation by opening up an entirely new area of potential interaction. The decision to push forward with the CFSP, while fitting in with the gradual extension of the European Union and the process of building a powerful, unified European voice on the world stage, is also a reflection of a pair of less grandiose desires. It is undoubtedly the inevitable consequence of European disillusionment with decades of exclusion from many of the decisions which shaped their immediate security, but it can also be considered a natural response to those in the United States who call for less involvement in the provision of security overseas.

That the system which helped provide security for Europe during the Cold War has gone is beyond debate. The alliance system which was created to service the Cold War purposes of containment is gradually being replaced by a more cosmopolitan and inclusive security construct. Within this broader international perspective sit the incremental moves towards the realisation of an ESDI, and the inclusion of the WEU as a partner organisation of the EU has clearly signalled the intention of the European partners to align themselves more closely together in matters of defence, security and political cooperation over these areas. While eventually it may have moved them closer to a re-Europeanisation of security, this is currently subsumed within a transatlantic framework.

This movement towards a Europe capable of managing its own security and defence affairs within a responsible partnership which includes the Americans and Canadians has only been aided by the resurgence in the activities of the WEU. One of the traditional European failings within the Atlantic alliance has been the level of reliance which was placed upon United States intelligence gathering and interpretation capabilities. While this is but one of the areas where the Americans bore the bulk of the Alliance burden, with more obvious areas such as logistical support and provision of the flexible deterrent force higher profile examples of European reliance on United States capabilities, it is one of the key features in any state or organisation's abilities to project its military power.

The interlocking nature of the organisational structure ensures that while the establishment of the WEU Satellite Data Interpretation centre at Torrejon in Spain has enabled the WEU and its member states to consider the possibility that the organisation can for the first time secure a level of independence in at least one of the fields within which it has had an established weakness, this need not be at the expense of NATO. Indeed an independent ability in this area will be crucial if the WEU is to take advantage of CJTF and further develop its operational capabilities. This clearly indicates that the desire to enable themselves to fulfil this role as a partner organisation is taking shape and thus one of the main weaknesses of the WEU is being addressed.

There does, however, remain a potential stumbling block to the success of this process of adaptation and it rests upon the American perception of the overall process. While the United States has in the past been a vociferous proponent of European independence in provision of its own security, and has increasingly shown a willingness to allow Europe to take control of its own problems when they arise, it is still a moot point whether they will be entirely happy with a Western European Union, which, either in isolation or in conjunction with the European Union, can assert itself on the world stage.

Even if the CJTF concept does allow for the United States to pick and choose when it becomes involved in military operations, it could prove the first step in a gradual dilution of American influence in Europe which does eventually lead to a solely European defence and security arrangement.

However, in the meantime there is still a strong transatlantic linkage, with NATO by no means ready to relinquish the position it has established as the foremost European security organisation. This fact is most accurately witnessed by evidence that the expansion of NATO is now a mere question of timing rather than a major issue of policy. The debate over the future form that an enlarged NATO will adopt is likely to rage on, but it is worth remembering that NATO is a political and a military organisation. As such there is more at stake as we discuss the Alliance's future than the stark issues of security co-operation. As the United States' main institutional linkage with the states of Europe it helps the United States maintain a priviliged position within the policy making forums of Europe and is thus unlikely to be relinquished

While circumstantial evidence points to the interpretation of the WEU as an alternative to the existing organisational structure, the sheer weight of logic outweighs this interpretation. Certainly in the medium term the United States must be involved in European security, both because they want to be and because the Europeans are still incapable of tackling the issue themselves. Europe itself still wants the United States on board and the French negotiations over re-entry into the military structures of NATO suggests that even they too, historically the most reluctant partners within the military framework of the alliance, are happy with the form of the new structures.

For now it certainly seems that the main themes of European security are likely to remain largely unchanged, with the enlargement of existing organisations and the consolidation of the working relationships between them all standing as the issues of primary concern to the members. Thus while the development of the WEU can be seen as providing some kind of potential alternative to the transatlantic framework, there is no immediate rush to pursue this and the majority of participants are happy to settle for the incremental, cooperative approach.

ENDNOTES

[208] Joffe, J., quoted in Conry, B., 'The WEU as NATO's Successor', *Policy Analysis No. 239*, http://www.cato.org/pa-239-html, (Cato Institute, 1995)

[209] Among the arguments developed in Conry, B., *The WEU as NATO's Successor*, Policy Analysis No. 239, http://www.cato.org/pa-239-html, (Cato Institute, 1995)

[210] Boutros-Ghali, B., *An Agenda for Peace: Preventive Diplomacy, Peacemaking and Peace-keeping*, Report of the Secretary General pursuant to the statement adopted by the Summit Meeting of the Security Council of the United Nations on 31 January 1992

[211] Boutros-Ghali, B., Ibid., Paragraph 65

[212] WEU Council of Ministers, *Petersberg Declaration, Part II On Strengthening WEU's Operational Role*, (WEU, Bonn, 19 June 1992), Paragraph 4

[213] WEU Council of Ministers, *European Security: a Common Concept of the 27 WEU Countries, Part I The Common Security Interests of Europe,* (WEU, Madrid, 14 November 1995), Paragraphs 36-64

[214] Assembly of the WEU, 31st Ordinary Session (Second Part), Report submitted on behalf of the General Affairs Committee, *WEU and the strategic defence initiative - The European pillar of the Atlantic Alliance*, Document 1034, (Assembly of WEU, Paris, 5 November, 1985)

[215] See the Rome Declaration on Peace and Cooperation issued by the Heads of State and Government following their summit meeting of the North Atlantic Council in Rome, 7th-8th November 1991

[216] See the Declaration on Western European Union, in the Final act of the Treaty on European Union, included as Appendix IV

[217] WEU Council of Ministers, *European Security: a Common Concept of the 27 WEU Countries, Part I The Common Security Interests of Europe*, (WEU, Madrid, 14 November 1995), Paragraph 13

[218] At the time of writing the French were still unhappy with the American unwillingness to free up the leadership of NATO's Southern Command to a European officer and were reluctant to commit to full reintegration without some indication that this attitude towards the Europeanisation of the command structure would change.

[219] This vision of the WEU as the defence component of the EU and as a means to strengthen the European pillar of the Atlantic Alliance is outlined in the Declaration on Western European Union, in the Final act of the Treaty on European Union, included as Appendix IV

[220] See for example the WEU Platform on European Security Interests signed at the Hague, 27 October 1987

[221] Cragg, A., 'The Combined Joint Task Force concept: a key component of the Alliance's adaptation', *NATO Review*, No.4, May 1996, pp.7-*10*

[222] This included twelve partner countries in the shape of Austria, the Czech Republic, Estonia, Finland, Hungary, Latvia, Lithuania, Poland, Romania, Russia, Sweden and the Ukraine as well

as troops from Egypt, Jordan, Malaysia and Morocco. Source: Admiral Leighton W. Smith Jr., 'The pillars of peace in Bosnia', *NATO Review*, No.4, May 1996, p14

223 Claes, W., NATO Secretary General, El Pais, 20 September 1995 - Taken from AM 259, DSC (95) 12, Draft General report from the General Rapporteur Rafael Estrella, October 1995

224 Solana, J., 'Shaping NATO for the 21st century', *NATO Review*, No.1, January 1997, p.3

225 The NACC had been created to provide a forum for the discussion of political and security issues. With the advent of the PfP, the security function has been somewhat superceded, although the NACC still forms a vital link to NATO for the majority of European states which are not members of the organisation. Its inaugural meeting was held on the 20th of December 1991, with 25 participants, including the 16 NATO members. By March 1996 there were 40 members of NACC.

226 The original batch of Associate Partners was limited to nine: Poland, Romania, Hungary, Bulgaria, Slovakia, the Czech Republic and the three Baltic Republics, Latvia, Lithuania and Estonia. This has been augmented in 1996 with the addition of Slovenia following completion of its 'Europe Agreements' with the EU.

227 WEU Council of Ministers, *Kirchberg Declaration*, 9 May 1994, Part II

228 WEU Council of Ministers, *Noordwijk Declaration*, 14 November 1994, Paragraph 10

229 For more detail on the activities of the NACC see NATO Fact Sheet Nr.1, http://www.nato.int/docu/facts/fs1.htm

230 WEU Council of Ministers, *Kirchberg Declaration*, 9 May 1994, Part II

231 Conry, B., *The WEU as NATO's Successor,* Policy Analysis No.239, http://www.cato.org/pa-239-html, (Cato Institute, 1995)

232 Mandelbaum, M., 'Don't Expand NATO', *Newsweek*, 23.12.96, p.17

233 Kinkel, K., 'The new NATO: Steps towards reform', *NATO Review*, No.3, May 1996, p10

234 Kinkel, K., Ibid., p12

235 Christopher, W., *A New Atlantic Community for the 21st Century*, Speech in Stuttgart, 6 September 1996, Washington File 723

236 Clinton, W.J., *State of the Union Address*, to a joint session of Congress (Washington, 4 February 1997)

237 Slocombe, W., *The Future of* NATO, Text of June 18th, 1993 Remarks to the Senate Committee on Armed Services, Subcommittee on Coalition Defense and Reinforcing Forces (United States Information Service, Washington DC, 6 July 1993)

238 The readiness of the USA to accept the concepts is developed by Salmon, T., in *Crisis or Opportunity? The Common Foreign and Security Policy and Defence*, a Discussion paper of the Jean Monnet Group of Experts, (Centre for European Studies, University of Hull), p.4

239 Millon, C., 'France and the renewal of the Atlantic Alliance', *NATO Review*, No.3, May 1996, p15

Chapter Seven

The Future Development of Military Space Activity in Europe

I. Introduction

As the millennium approaches, the countries of Western Europe are having to make some serious choices as to the overall development of the security architecture which will endure into the next century - a topic which will be more fully addressed in the next chapter. However, one important element which figures in such planning is the role which space will play in such a framework.

Consequently, having examined the growth of interest in this area, and the specific role of the WEU in such developments, we must consider the way in which the European use of military space may develop in the future. If, as has been suggested, this is an increasingly vital element of any security or defence strategy, how are the states of Western Europe going to take forward their, thus far, limited involvement in this area?

II. National Aspirations

Historically, France and the UK have been the only Western European states which had any real autonomous military space capability, through their respective Syracuse and Skynet programmes. More recently, however, this has been changing. In the 1990s, as we have seen, other countries including Spain, Italy and most recently Germany, all have some military space systems, or at least well-advanced plans for them. The overwhelming majority of these systems, almost all of which are communications systems, have been developed and deployed by individual states. It is logical to assume that the number of European states involved in such programmes has increased because each has come to the conclusion that, in an overall assessment of their operational requirements, such systems provide an efficient and cost-effective means of meeting those requirements.

It is also reasonable to assume that demand for the type of capabilities offered by such satellite systems will continue to grow. It is interesting to note that no state, having gone down the road of developing and deploying a military communications capability by way of space-based assets has ever abandoned this capability (although the British did consider this option in the 1970s).[240] Even if this trend were to be bucked and a state did, at some indeterminate future time, decide it no longer needed military communication satellites, it must be considered most unlikely that this would ever apply to more than a mere handful of states. Thus, the overall trend will be an increasing, and continuing demand for such space systems.

The multi-million dollar question arising from such an assumption is whether or not the states involved or currently contemplating involvement in the exploitation of military space assets, will continue to develop such systems on a purely national basis or will instead seek to gain the economic efficiencies afforded by cooperation or collaboration in their development and deployment. All the evidence at this time tends to suggest that the latter option is the more likely, with the overall cost of space based systems such that, at a time of increasing pressure on defence budgets, any savings which can be obtained would be wholeheartedly welcomed by the accountants, a breed of whom seem to have growing influence in defence planning matters.

This generates a number of supplementary questions:

- how might such cooperation be organised?
- will it be done at the programme level, as has been the case in the few examples of such cooperation so far?
- will a more radical path be followed, with military space activities coming directly under the auspices of an international organisation?
- if the latter option is developed, which organisation would be most suitable?
- is this a role which could viably be assumed by any existing international organisation or might it require the establishment of an entirely new organisation?

All of these questions remain to be answered if the exploitation of military space assets is to become a feature of the increasingly robust nature of the European approach to matters of defence and security.

Chapter Five examined in some detail current European military space capabilities. However, it must be borne in mind that many of the projects so far undertaken are largely national in nature. In those instances where there has been multi-national cooperation, such as the Helios programme, it is fair to say that one country has very largely been the driving force behind the programme with the involvement of other states comparatively limited (in the case of the Helios programme the primary participant was France). Will this continue to be the pattern for future military space activity, or will the individual states come to the conclusion that it might be more logical to pool their efforts and resources in a truly pan-European enterprise?

It took many years for such a path to be followed in the realms of civilian space endeavours. The net result was the very largely successful European Space Agency (ESA), which is now undoubtedly one of the major players in the international space arena.[241] Is it then possible that ESA could serve as a model for military space activity as well, or are there special circumstances relating to military cooperation which may mean that such an outcome is unlikely?

III. Collaboration at the Programme Level

Technological Synchronicity

The most basic requirement which must be met to allow any real collaboration at the level of individual programmes is that all the countries which may be considered potential collaborators must perceive a similar technological demand. In other words, they must all be seeking a broadly similar kind of service from the proposed system. While this may seem self-evident, it is nonetheless a potentially major stumbling block. Such questions featured in the discussions held on the possible INMILSAT system, between the US, the UK and the French where the Americans and the British, having different requirements, desired different frequency transponders designed to serve different purposes. While it could not be suggested that this was the only reason for the failure of this project to get off the ground, it was certainly one of the more problematic elements of the negotiation process. Thus it can be seen that different countries can and do seek different capabilities. During the INMILSAT study, which was originated by the French in 1991, several different design options were considered. Essentially the Americans preferred a purely SHF system, while the British and the French wanted a combination of EHF and SHF.[242] While one possible answer to such a discrepancy would be to install a variety of different transponders on the satellite, this may not always result in a satisfactory range of capability for each of the participating states, with the major technical limitation of size ensuring that there is a finite number of transponder packages which can be mounted on one satellite bus. In such a situation, the possible loss of capability for one country may be regarded as an excessively high price to pay, with the price not only measured in financial terms, but also in terms of the operational capability that any system offers.

Similarly, different participants may well have different requirements in terms of the geographical area of coverage that they need. One may desire wide area coverage of a specific geographical area, while another may prefer to have access to steerable spot beam antennae which provides a more specific localised coverage. While such discrepancies may not be insurmountable, they again must result in compromise and thus a potential degradation of capability for each country involved. The acceptability of such a reduction in specific capability must be measured against the economic advantages of collaboration, and indeed the political advantages or disadvantages involved.

In outlining current and planned military space programmes in Chapter Five, with reference, for example to the TRIMILSATCOM programme, it was pointed out that one of the key factors in facilitating cooperation in a given programme was synchronicty of requirement. If one takes, for example, the potential for collaboration in such a system, as proposed, between the British, the French and the Germans, both the UK and France have existing military communication satellite systems, respectively Skynet 4 and Syracuse 2. The potential for collaboration in building a joint successor system is considerably enhanced if both the current national systems will reach the end of their useful life at the same time (a fact which can only be estimated roughly as the precise dates on which such systems are expected to cease to be operationally useful is classified). Let us assume, however, for argument's sake, that the UK Skynet system will reach the end of its useful lifespan, say, two years before the French Syracuse system. When should the joint successor system be launched? Presumably the UK would insist that it should be operational before their current system ceased to be effective, otherwise they would face a period of time within which they would have no military communications satellite capability - a situation which the Ministry of Defence would presumably find unacceptable. The French on the other hand would presumably be unwilling to launch their successor system two years before it was actually required sincc they would then be in a position of having 'wasted' two very expensive years of satellite operation.

It is, of course, true that such a situation does not necessarily produce a totally insurmountable problem. Some form of compensation to the state 'wasting' their satellites could be built into the collaborative agreement in the first place. Nevertheless it can be seen that synchronicity of demand is an important element in determining whether such collaboration is possible. If the difference in lifespan of existing systems is relatively small, perhaps a year or so, then collaboration may be appropriate and possible. If on the other hand the difference was a period of three or four years, then this might indeed endanger the whole prospect of collaboration. Thus, to some extent, there is an element of chance involved in the determination as to whether, at the programme level, cooperation or collaboration is viable. In relation to this consideration, Germany has, of course, been omitted since such a problem does not arise insofar as they do not currently operate a military communications satellite system.

There are two other elements which must be taken into account in assessing the viability of collaboration at the level of individual programmes or projects - the economic and political implications.

Economic Criterion

Given the very large financial commitment required in any such programme, each potential participant country must be convinced that the proposed collaborative programme will, at the end of the day, provide a level of operational capability which is acceptable to that state, at a lower overall budgetary cost than would be the case if a purely national system were to be developed instead. No-one would dispute all space systems are expensive, and thus the potential for cost-saving is one of the strongest motivating factors in raising the issue of collaboration. If one looks at the EUMILSAT proposal as an example, the estimated figures for the space section of this programme were projected at up to $1.38 billion. The cost of a typical national system is around $600 million (Syracuse 2).[243] Thus it can be seen that the overall cost of the collaborative project is roughly double that of a national system. If, however, the cost of the collaborative system is shared amongst three members, the cost to each will be proportionately less expensive.

However, there are also costs involved in collaboration. The greatest single such factor is that the management overheads of the project will inevitably be increased. Perhaps the best way of illustrating such increased costs is to highlight several very simple examples. At the planning stage, and indeed throughout the project, there will be a need for consultation and discussion between the relevant directing agencies of all the countries involved. This process will, of course, take place in a number of different ways, ranging from the exchange of letters, through telephone conversations to visits to one's counterparts in the other participating countries for face to face discussions on particular aspects of the project. All these cost money. Another simple example which can be cited is that of the project documentation. Clearly, not only will it be necessary to produce additional copies for those directly involved in the project management in each of the countries, but it will very likely be necessary for such documents to be translated into one or more languages other than the one in which they were originally written. Again, this takes time and costs money.

While such examples may seem trivial, particularly when set against the likely multi-million dollar cost of the project as a whole, they serve to illustrate the ways in which the management costs of any programme are driven up by international collaboration and these problems are magnified as the number of countries involved is increased. In a purely national project the decision to change one relatively small component of the overall system may be almost entirely insignificant. If such decisions require to be approved not by one national procurement agency, but by two, three, six, or more, such apparently minor decisions can consume considerable quantities of time and, once again, can prove to have considerable resource implications.

In general terms, if such problems can be caused by relatively trifling design changes, imagine the difficulties brought about by major changes, or the immense problems that can be caused during the original definition and design phase of the project requirement. To cite one (non-space) example, this is effectively what brought about the demise of the NATO Frigate (NFR 90) Project in the 1980s. This eight-nation programme, which would have resulted in almost 70 ships being built, started in 1984-85 with a feasibility study, but serious differences emerged between the participant countries over the basic size of the proposed vessel. This, along with cost over-runs, led to the UK withdrawing from the project at the end of September 1989, and this was followed just a few days later by announcements from both France and Italy that they too were pulling out of the programme. This serves to highlight the difficulties inherent in major collaborative programmes. After more than four years, the nations involved could still not reach an agreement on such a basic requirement as the size of the vessel, or settle on a realistic timetable for the production of the vessel itself and its weapon systems.

If two countries collaborate on a programme, they cannot expect the overall savings to each to be in the region of 50 per cent, the saving to each may be more in the region of 30 per cent - still, of course, a substantial saving. Against this, however, may be counted some reduction in operational capability, and also the need to share decision-making and control with partner countries.

Political Considerations

Here we move into the territory of the final factor to be considered - the political dimension. Military space systems are, of course, part of the overall security and defence structure of the state. Defence is perhaps one of the single most important priorities of any state - indeed from the days of the Treaty of Westphalia, territorial security has generally been regarded as the *raison d'être* of the state itself, the highest of the issues considered 'High' politics. It is therefore a considerable leap of political faith to cede any responsibility for this crucial area to allies. Such arrangements can, from time to time, work. In one sense the whole framework of the North Atlantic Treaty gives an example of such a leap of faith. It established a multinational collective defence organisation which, for some 50 years, has provided the bedrock of European and transatlantic security. Clearly, such arrangements have been beneficial to all the concerned parties throughout that time, and thus cooperation in matters of national defence can be successful.

Despite such grounds for optimism, collaboration at the programme level, at least in terms of military space systems, is a rather different matter altogether, and raises a number of different questions. NATO is principally a political alliance which operates on the basis of consensus amongst all its members, normally within the North Atlantic Council. In this sense, therefore, each and every NATO member state has a right of veto over policy decisions. While the vast majority of NATO member states also choose to participate in the integrated military command structure of the organisation, this is not a pre-requisite and of course there are several who have chosen not to participate in this way, the most notable being France.

However, it is inconceivable that such a system of operation, reliant on the principles of debate and consensus, could operate at the level of individual programmes in relation to military space. A military space system is an operational military facility which must be run, round the clock, every day of the year, as an operational facility. Thus, decisions as to its use must be taken in a timely and coherent manner. There must be a clear chain of command in terms of its operation at all times. Where such a system is operated jointly by a number of countries, be it two, three or even more, one has to ask the question as to who is in overall command of the system. While on a day-to-day basis this may not pose any great problem, in a time of crisis it is not difficult to envisage a degree of conflict arising between the operating states, each of whom may - indeed most likely will - have differing needs for the system at a given moment. In such a case, who takes priority? There may well not be time for a meeting to thrash out priorities between several different states and arrive at a consensus. In this scenario therefore, one state, no doubt through the senior military officer for the programme, will effectively impose their will, or that of their national authority, on the use of the programme. How willingly would other countries accept such a scenario?

In a sense this raises a somewhat broader, more general, political question. The comparison with NATO is not wholly accurate, in that NATO does not, to any significant extent, replace national responsibilities for security and defence. It adds a further layer of defence cooperation which only truly comes into effect in the particular circumstance of an attack on a member state. All other planning, cooperation, and discussion in relation to other matters is, in the strictest of senses, a side-show. No decision taken within the overall framework of NATO impinges to any great extent on the national defence and security policies of its member states. True, NATO offers a mechanism through which matters pertaining to national security positions may be discussed and agreed - force levels and roles, for example. But, as has already been pointed out, such decisions require unanimity - no individual member state can be forced into roles or decisions which their national security perspective fundamentally disagrees with. However, in a purely collaborative programme, no such guarantees are necessarily available to participating states.

This fear notwithstanding, the political dimension need not always be regarded as a negative factor. The other side of the coin is of course that collaboration may be deemed politically advantageous because of the overall integrative role which it serves. Just as the economic rationale suggests cost-sharing, resulting in cost-saving, in political terms collaboration may be seen as a positive act when it serves to bind together two or more countries. The burden-sharing debate which ran for so long within NATO was as much a question of political roles as an economic argument. Thus, in attempting to further political unity, or political integration, military space may simply be seen as another functional area within which such steps may meet with success. To this end, therefore, there may be very positive political motivating factors, even at the programme level.

In this regard we need look no further than the Maastricht Treaty on European Union, which clearly states that:

> '... the common foreign and security policy shall include all questions related to the security of the Union, including the eventual framing of a common defence policy which might in turn lead to a common defence'.[244]

Following on from the logic of the Treaty, the question of military space policy may legitimately be regarded as but one element relating to the security of the Union. Thus, any move towards a collaborative military space policy may be seen as a positive step in the spirit of the Maastricht Treaty.

It is of course possible that some form of collaboration could be undertaken without necessarily any significant level of integration. When the EUMILSAT proposal was first floated in 1991, with the UK and France seeking further international partners, the programme was discussed at the technical level, a deliberate decision designed to avoid politicisation of the process at that early stage. Michel Javelot, head of telecommunications at the French Electronics Directorate, pointed out that 'a transmission capability that is shared between several European countries in no way implies any military integration. EUMILSATCOM can be considered as an agency that sells communication capacity, like a telephone company, and nothing more.' 245

This form of minimalist approach appears attractive on the face of it but may be seen to have a number of drawbacks attached. While such purely inter-governmental cooperation as an international agency might work in normal circumstances, it must be remembered that military communications systems are specifically required to work in abnormal circumstances. Following M. Javelot's analogy of a telephone company, when the lines are busy, the resulting engaged tone is annoying. In the realm of a military communications system, however, the consequences could be disastrous. With such an agency, who would dictate priorities if two countries are involved in different operations, in different geographical regions? Such an eventuality would require significant and competing use of the satellite system. Unlikely though such a scenario might be, the art of military planning is to endeavour to plan for all eventualities. It is thus difficult to envisage how such a multinational system would operate without direct political and strategic control. Indeed, such considerations must be regarded as among the most compelling reasons why such an idea stalled.

IV. Higher Level Cooperation

So far we have considered only the potential risks and benefits of collaboration at the programmatic level. It is of course entirely possible to envisage collaboration or cooperation at a rather higher level - a truly intergovernmental, multinational arrangement. Indeed it is possible to argue that the WEU Satellite Centre in at least one sense fulfils such a role. The WEU, as an organisation, clearly meets all the necessary criteria. Thus, the Satellite Centre, as a subsidiary body of that organisation, has identical membership of states and so on. In that sense, therefore a multinational, intergovernmental military space body does already exist.

In the broadest sense, the governments of Western Europe already support the concept of collaboration and cooperation in the field of defence procurement. Whether independently, through bilateral or multilateral agreements, or through the overall framework of NATO, there is at least a nominal commitment to explore collaborative ventures. To date, there has been little evidence of this in the specific area of military space systems, Helios being the notable exception to this statement. However, today, with Helios 2, Horus, and the possibility of TRIMILSATCOM, at least at the independent, programme level, there seems to be greater enthusiasm for collaboration than has been the case.

Is it possible, therefore, that the spirit of the times is changing - that such enthusiasm may transcend the individual programme level and lead to even greater international cooperation? The rapidly and radically changed political and security environment which has emerged over the last 10 years has forced defence planners to re-examine completely their force structures and also their budgets. While the traditional primary threat to Western European security has, for the present at least, evaporated, a much wider range of security scenarios must now be considered and planned for. In the NATO parlance of the early 1990s, the single specific threat has disappeared, and we are left with a much wider range of unspecified risks. This has, however, caused considerable pressure on defence budgets as a whole, which has in turn resulted in an even greater need to justify potential spending on new and expensive military projects. As we have seen, the development of space-based systems is one of the most expensive areas, and thus perhaps one of the most difficult to justify.

That said, it is still possible to turn the argument around, and to contend that recent political and strategic events actually make it easier to argue a case for the development of a unified European military space policy. In order to better meet the wider range of risks which Europe faces, it must rely on the best possible sources of intelligence, which should ideally be independent and European, thus diminishing reliance on other states. Additionally, it is not difficult to envisage scenarios in which versatile, reliable, and again autonomous, wide-ranging communications capabilities may be necessary if the states of Europe are to ensure their capability of playing a full part in the new security arena. Such capabilities can, in many cases, be met most efficiently by military space systems. In cases such as surveillance and communications, the high absolute cost of space-based systems is offset by the fact that, in relative terms, they still offer considerable value for money. In such cases it is possible to make the case that they are cheaper or better (or perhaps both) than the alternatives. Furthermore, such economic pressures are likely to promote the concept of international collaboration.

V. The Civilian Example

The European space industry, at least on the civilian side, has a comparatively lengthy, and very largely successful, history of international cooperation and collaboration. The European Space Agency, if not a wholly perfect example of European cooperation has, nonetheless worked reasonably well in its 25 years of existence. It has permitted the nations involved to carry out an overall programme of space research and activity which would have been entirely unattainable for any individual European state. While it has inevitably faced political difficulties from time to time, its future seems guaranteed, at least so far as anything can be certain in an uncertain and constantly changing world. There have been times when the intransigence of one or more member states has caused difficulties. An example of this was the position of the British in 1987, when they declined to participate in any of the projects mooted at the Ministerial Council in the Hague. More dangerously for the organisation as a whole was their reluctance to agree an increase in the Horizon 2000 science research programme, one of the core, mandatory elements of ESA. Eventually, some months later, the UK government reversed its position and did agree to the 5 per cent increase in Horizon 2000 funding, thus avoiding a potential crisis which could ultimately have resulted in the UK being forced to leave the ESA altogether.

In general, however, such intransigence has, to a very considerable degree, been offset by the enthusiasm of others, and, although it does continue to run into the odd crisis over specific programmes, or more commonly, over funding, it does, in the main, manage to overcome these potential stumbling blocks and provides a useful focus for a concerted, reasonably coherent, European civilian space policy. Clearly the situation could be better. In an ideal world, each nation would agree that the overall European aims and objectives should take priority over their own, rather more parochial, national policy. At the same time, in this mythical ideal world, there would be a marked unanimity of purpose, a general agreement of space research and development, and the whole concept would be carried forward in a glorious spirit of supra-national harmony.

Unfortunately the real world is not quite like this. Members of the space community must accept the harsh realities of life: such harmony and unity of purpose are at the moment largely pipe-dreams which may be sought but are most unlikely to be achieved in the near or medium term. Economic and political pressures create an environment in which the wisdom of investing comparatively large sums of money in the field of space activity is constantly being questioned. It is an environment which, while not in theory opposed to international collaboration, remains downright suspicious of its benefits, and in which national needs and aspirations remain paramount. Nonetheless, ESA, despite having to operate in this far from ideal world has produced concrete results and shown that cooperation, at an agency level, in the highly competitive arena of civilian space not only can work but does work. This of course gives rise to the obvious question - would such an agency be appropriate for military space activity as well?

VI. A European Military Space Agency

In considering the viability of a European agency to undertake activities in the military space arena there are several institutional arrangements already in existence which provide realistic options:

European Space Agency

There are major difficulties in utilising the existing institutional arrangement of the European Space Agency for undertaking a military role. Article II of the Convention of ESA expressly states that its purpose is to 'provide for and promote, for exclusively peaceful purposes, cooperation among European states in space research and technology.' The key phrase in this context is, of course, 'for exclusively peaceful purposes'. This sentiment can be clearly traced back to the very origins of the European space programme, when the founding fathers were adamant that there should be no connection with the military at all. Edoardo Amaldi, for example, one of the original guiding lights of European space organisation was passionate in his belief that that military had no part to play in such an endeavour.

> 'Amaldi's project [the establishment of a European space research organisation], in fact, was based on one main pillar: the military were to be kept out of the future European space organisation. Firstly, and this was the "moral" side of the coin, he conceived the conquest of space as a peaceful enterprise open to all mankind; therefore it should not remain an exclusive property of the military of both fields. Secondly, and this was the "practical" side, he considered the military environment as an obstacle to the best development of science, characterised by open co-operation and free exchange of ideas.'[246]

Such sentiments, have, since that day, been espoused in the various European space organisations which have developed, right up to the present day with ESA. The result of this institutional attitude is a clear limitation on the extent to which ESA may be involved in any military programme. This does not totally preclude all ESA involvement in the military uses of space, albeit that such involvement has been largely indirect and involuntary. The most obvious example here, is the Ariane launch vehicle, which has been designed and developed by ESA, and which has already launched military communications satellites as part of the package of commercial launches which have over the years helped establish Ariane as one of the most reliable names in the satellite launch business.

While such involvement can hardly be seen as direct ESA involvement, the satellites are after all not their concern, it would almost certainly be possible, albeit with considerable debate, to arrive at the conclusion that such a phrase need not preclude ESA from being a major participant in, for example, a satellite system whose purpose was to verify arms control agreements. It is fairly easy to see how this could be interpreted as a wholly peaceful purpose, which would therefore not be at odds with either the text or the spirit of the ESA Convention. In such a case, it must be conceded, however, that it would be difficult, if not entirely impossible to prevent states from using the data obtained from such a system for rather more aggressive purposes, such as, for example, targeting. Any system which is capable of ensuring compliance with existing arms control agreements would almost certainly also have the capability of providing its operators with useful military intelligence, which could be utilised in more general defence planning.

Similar arguments could be made with respect to other fields of operation, such as crisis management. Again, the necessary data for such peace support operations, which could fall within the remit outlined in the ESA Convention, could also be utilised for purposes which might not be considered wholly peaceful and which might then damage the credibility of ESA's non-military ethos. This paradox lies at the heart of any argument which suggests that ESA, as it is currently structured, could act as the appropriate agency for the conduct of any serious military space policy.

It is also very likely that such a system would be only one element of a more varied overall military space capability. In all likelihood, a concerted European military space policy, if such a thing was to be established, would include other such elements as command, control and communications facilities, as well as more general reconnaissance and surveillance capabilities. There is also always the risk, in planning terms, that at some indeterminate time in the future, perhaps 20 or 50 years hence, it may be deemed necessary or desirable to develop and deploy rather more offensive space assets, such as an Anti-Satellite (ASAT) system, for example.

It therefore seems that to base a limited, short-term military space policy on an agency which was thus restricted to purely peaceful objectives would be generally unwise. Although it is quite possible that it could fulfil current, or even medium term requirements, it would, in all probability be too restrictive to allow for possible, currently unforeseen future developments.[247]

North Atlantic Treaty Organisation

The most immediate inherent problem in suggesting that NATO could offer the institutional framework for a European military space agency is its composition. If the objective is to establish an autonomous European programme, the presence of the United States and Canada poses a fundamental problem. As has already been indicated, one of the fundamental reasons why Europe is, and arguably should be, considering the relatively expensive acquisition of autonomy in this area is the realisation that there will be times when North American and European security interests do not necessarily lie in precisely the same plane. There have been, and doubtless will be again, times when one particular area of the globe is of immediate interest to the European nations, but not to the Americans or Canadians. Indeed, if this was not the case, there would be no need whatsoever even to consider enhancing Europe's independent military space capabilities. The mere fact of alliance with its transatlantic partners would always suffice to meet European military space requirements.

An aspect which further limits the possibility of any real role in this respect for NATO is the whole question of NATO enlargement. If the current membership raises questions in relation to a European military space policy, this problem can only be exacerbated as the process of NATO expansion eastward is put in train over the next few years. One of the inevitable side-effects of NATO enlargement will be a dilution of the effectiveness of the decision-making process. It will, in such a case, be more difficult to operate, on a day-to-day basis, multinational systems - a price which has to be paid for the political benefits of enlargement.

There is no doubt that NATO has served as the principal actor in both transatlantic and European defence policy for the last 40 years. Despite the changing international security environment, which may well result in further major changes in the overall security architecture of Europe this situation is likely to persist. However, because of its composition, it is not really feasible to consider NATO as the appropriate agency to operate European military space policy. While there can be little doubt that it will work closely with whatever agency may take on the primary responsibility for such policy, it could not, realistically, take on that role itself.

Western European Union

As far as existing European international organisations with a security remit, and, most significantly, some form of operational capability, is concerned, this leaves only the Western European Union as a possible provider of the institutional framework necessary to develop, and operate a Western European military space policy. At first sight, given the evidence of interest already covered in some detail, there is much to commend it as the appropriate body for this task. It is composed entirely of European states, and so meets the criterion of European autonomy. One of its principal functions, as it is presently constituted is the consideration of a distinctively European view on defence and security matters. These are relatively strong credentials, compared with other possible existing international organisations.

However, despite these undeniably strong arguments in favour of the WEU, it may not be the ideal body either. The main reason for this rather negative contention lies in the relative uncertainty as to its future role. To date, examples of WEU operational capability, such as the Satellite Centre and the role it played in the Yugoslavian crisis through operations such as Sharp Guard, have all been rather limited. While it is true that a more dynamic, ongoing operational capability could no doubt be developed over a period of time, as it currently stands, the WEU is a relatively small organisation, just about capable of the limited operations it has been involved in. Politically, it is possible that other organisations, such as NATO may not be keen to see this relatively junior organisation expand into significant operational areas. There is also the question of the relationship between the WEU and the EU, as originally espoused in the Maastricht Treaty on European Union. Again, if such a path were followed, the combination of WEU offering a security perspective to the EU and an enhanced operational capability for the Brussels Treaty organisation may also be regarded as something of a threat by the 'big brother' of NATO. Whether this would at the end of the day be acceptable is a question which only time - and no doubt considerable diplomatic negotiation, will answer.

There is also the question of membership. This already poses some difficulty in that the memberships of the EU and WEU, while overlapping to a considerable extent, are not identical. As in the case of NATO enlargement, this situation may be worsened by further EU expansion, unless WEU membership comes along as part of a package deal - a situation which is by no means certain to develop. There is also another factor, which considers the possibility that perhaps not all the members of WEU necessarily regard it as being a strong enough entity to develop an on-going operational capability. Nonetheless, the fact remains that of all the existing European international bodies, the WEU has the strongest credentials to take on the responsibility of a European military space policy - and the operational applications of such a policy.

But if it is still not the ideal agency, and for a moment one assumes that a concerted European military space policy is something to which the European states aspire, is there then a case for the establishment of a wholly new body - a European Military Space Agency in its own right?

A European Military Space Agency

If, as suggested, none of the existing international organisations in Europe are particularly suitable for the determination, development and operation of a European military space policy, would an entirely new organisation be any better placed to fulfil this role. Indeed, at a time when the overall security architecture of Europe is still under active consideration; when existing organisations are re-evaluating their roles and functions; and when new inter-relationships between existing organisations are also being formed, when better to consider the establishment of a new regime. Might this not be a good time to go for an innovative policy move when existing regimes and policies are moving in perhaps radical directions anyway?

It may, at the very least, be an appropriate time for the governments involved to include in their overall calculations and deliberations on the future organisation of European security, an assessment of the role which they see space playing in the security of Europe. Increasingly they are taking steps to examine or develop military space systems. They must now evaluate whether it is more satisfactory, more efficient, and will lead to an enhancement of European security, if such activities are carried out in a cooperative or collaborative manner with their European allies.

There are a wide range of factors involved in such a prospect, including all those already identified in relation to other scenarios. Such a decision will not, and indeed should not be taken lightly. If, however, at the end of the day they were to decide that such a step - the establishment of a European military space agency (EMSA) - was appropriate, then what format might it take?

In effect, all the necessary considerations boil down to a question of control - political, operational, and economic. Of these three, the most important is the political element, for without adequate arrangements for political control, including overall policy and decision-making, which are acceptable to all the states involved, then there is very little chance of success for the venture. Perhaps the most obvious way to consider the form and structure which such an agency might take is to look at the necessary component parts, and quite unashamedly, borrow the best, most relevant elements from existing organisations, to give an overall pattern which might be followed.

At the political level, the key element is inclusiveness. Each of the potential member states must feel that they have a sufficiently significant voice within whatever decision-making processes are developed to ensure that, at all times, their national interests will not be subsumed to the European interest. For as has been seen in every other instance of a European organisation, a significant number of states are not yet prepared to subjugate their national interests. To allow for this, it must be regarded as almost essential that policy decisions are taken on the basis of consensus. It is inconceivable that countries such as the UK, or even France, would permit a situation to arise in which at least one aspect of their national security policy could be dictated to them by a majority vote in such an institution. Consequently one can envisage, much along the lines of the North Atlantic Council, an EMSA Council, within which each state is represented, and within which policy decisions can only be made on the basis of consensus. While this would meet regularly at diplomatic level, it would also be able to meet at ministerial, or Head of Government level from time to time. At

Ministerial level, however, one might suggest adopting the WEU Council practice, of having both the Defence and Foreign Ministers of each country on the Council. There is little doubt that this is one of the distinctive elements of the WEU which has, over the past 13 years or so, added to its effectiveness in considering security matters.

Operational control is another aspect which, on the face of it, poses a number of problems. For effective operational control, there must be a clear chain of command, culminating in one individual in the overall leadership role. While this may be problematic, in terms of the nationals of one state being subordinate to another, this is not insurmountable. Nor, of course, would it be unique to this organisation. Such a situation already exists in a variety of international organisations - indeed, in the case of all operational organisations, whether military or otherwise, it is effectively the norm. It can, from time to time cause difficulties, as it has within, for example, NATO. However, assuming the political will exists to ensure the fundamental strength of the organisation, and this is shared by all parties, such issues are generally able to be resolved. The question is, how best to suggest resolution?

One might consider that one should again follow the NATO pattern. Given that EMSA would be, as the very name suggests, a military agency, it may be felt that the military authorities from each of the member states should have direct input. In such a case one could see a need for a Military Committee (NATO-style) comprising the Chiefs of the Defence Staffs of the member states. However, given the overall remit and size of the Agency, which would of course be very considerably narrower and smaller than that of NATO itself, this might be regarded as an unnecessarily cumbersome structure. However, it is not realistic to suggest that there should not be some form of military input to the Agency. One solution may be that the Chiefs of Staff offer their advice at the national level, to their respective Ministers, and that the direct military input to EMSA be at a somewhat lower level - at the level of those involved in the Agency on an operational basis, for again, it is inconceivable that such a military agency will not be staffed, at least to some extent by serving military personnel.

Which leads to a further operational question. Should the staff of such an Agency be predominantly civilian or military personnel? The very title of

Agency tends to suggest that the answer to this question is civilian. This may also help to resolve the preceding question of military representation. There is no reason why the staff of such an organisation must inherently be serving military personnel. There are numerous examples of security agencies, or other bodies which provide intelligence and support to the military which are staffed by civilians. In the UK for example, bodies such as the Government Communications HQ (GCHQ), or the Defence Intelligence Staff, are essentially civilian bodies. There is therefore no reason why this pattern could not be adopted by EMSA. If this route were followed, then each member state could appoint one military liaison officer to the Agency. These individuals would individually provide national military input to the Agency and its work, and collectively form a Military Committee which could advise the Council and the Director-General as required.

For if this is the model to be adopted, one clearly requires someone at the operational head of the Agency, and, if it is to be essentially a civilian body, the title of Director-General would seem most appropriate. This individual would be appointed on a fixed term of office, perhaps three or four years, by the EMSA Council, and, as in the case of ESA, or the existing WEU Satellite Centre, would provide the day-to-day leadership, with the responsibility of carrying out the missions and tasks assigned by the Council. At the more detailed level, he or she would doubtless be supported by one or more deputies or assistants, and the running of the Agency would be devolved either through divisions or departments. Clearly, consideration would also have to be given to the lowest level at which appointments would be the responsibility of the Council, but one may presume that their control would extend down to at least Assistant Director-General.

The third question of control which was posed above was that of economic. This is perhaps not as difficult as might originally be thought. If a structure, not dissimilar to that which has just been outlined, is adopted, since the Ministerial responsibility for the Agency at national level would be held jointly between the Defence and Foreign ministers, it is reasonable to assume that the financial burden for the agency would also be shared between these two ministries in each of the member states. The precise level of funding to be obtained from each state would be determined annually by the Council - a common arrangement in international organisations. However, it must be remembered that disagreements over funding are also one of the most common difficulties encountered by international organisations. For the system to work effectively, each state must feel that it is not being asked to shoulder an unfair proportion of the budget. One need look no further than the United Nations, and the difficulties they continually face with contributions, as an example of such problems.

Again, the solution lies close at hand. In this respect, the funding basis of ESA - which was, after all, devised precisely in order to deal with such difficulties - would seem to be the most appropriate model. In the ESA case, the budget is divided into two distinct components - a mandatory, core budget, and a series of optional programme budgets. All member states must contribute to the core budget, at an agreed level, in order to maintain their membership of the organisation. This budget covers all the central functions of the organisation - and in the ESA case the basic space science activities as well. Other programmes which may be undertaken by ESA are however optional and it is up to each individual state to determine whether or not they wish to participate, and at what financial level. This would seem to be the ideal funding basis for EMSA as well.

Again, the core activities of the Agency, at a fairly modest operational level, would form one budget, with each member state being obliged to contribute a predetermined proportion of this budget. This would ensure general financial stability for the organisation as a whole. Other elements of its activities, such as the development of new satellite programmes, or specific tasks, might be designated as options, thus allowing for states to decide whether to participate or not. One amendment which should be considered is the addition of a compulsory contingency fund, to which all would contribute. This fund would provide an element of financial security in the event of a crisis suddenly developing, as a result of which specific operations or tasks were assigned to the Agency at short notice. This would mean that militarily important tasks could be immediately undertaken without there having to be any real debate as to the attribution of costs in such a case.

VI. Conclusions

This chapter has attempted to examine some of the potential problems inherent in the future development of a coherent expansion of European military space policy. To date, there has been relatively little significant progress in this area. However, in different ways, the groundwork has been laid. The European Space Agency has demonstrated that cooperation within the general field of space activity is feasible at the European level. Indeed some might go further and argue that not only is it feasible, cooperative endeavours are the only cost-effective way forward in this area. Space activity is just so expensive that no single European state can afford to go it alone.

There is also a long and honourable tradition of cooperation in the field of security or defence as illustrated by the general success of the NATO framework, which has undoubtedly played a most significant role in ensuring the security of Western Europe over many decades. However, as the overall security requirements of the continent change, so too must the institutional framework which is designed to ensure security. Thus, a growing desire for Europe to engage more fully with its American cousins on a more equal footing has emerged. There is also a growing desire to see a greater degree of European autonomy. Following on from this, there is, at least in some quarters, a perceived need for Europe to acquire capabilities in areas which hitherto have been largely the domain of the USA. Military space capabilities are one such area.

There have undoubtedly been some steps in this direction. The establishment of the WEU satellite centre is a landmark in such European cooperation. Similarly, the launch of multi-national systems, such as Helios, or the proposed TRIMILSATCOM are also evidence of a willingness to cooperate in this field. The significant question that remains to be answered is whether comparatively piecemeal developments among a variety of partnerships of European states are the most sensible way to proceed. Would it not be more efficient, more cost-effective and more productive for the European states interested in such developments to work together more closely, eventually leading to the production of an overall military space policy for Europe? If that is indeed seen as the most sensible way to approach the matter, then consideration has to be given, at the earliest stage of planning, to how such a policy might best be implemented.

There seems little doubt that the only organisation currently in existence, which could assume the mantle of Europe's military space co-ordinator, is the WEU. In essence therefore, the real choice lies between that body or an entirely new military space agency.

In the final analysis, the new agency may appear less than desirable for the simple reason that the further proliferation of international organisations with a security remit would be less than helpful, only serving to further confuse the already crowded organisational landscape. Thus the most sensible option would seem to be the development of military space policy within the WEU framework. This could be achieved by extending the role of the non-permanent Satellite Centre, or alternatively, creating a further WEU subsidiary body, a military space agency which could be structured along the lines suggested above. For the reasons discussed in this chapter, this appears to be the most pragmatic, sensible and perhaps only realistic approach.

This is the area which this chapter has sought to address. Some may regard the arguments against one or more of the existing international organisations in Europe as not necessarily rendering them incapable of carrying out this new role. Others might contend that the outline of a possible European Military Space Agency is flawed in some respect. We make no claims to offer the perfect solution, or an answer to any of the many difficulties which would undoubtedly be involved in such a step. Rather, this is an attempt to open the matter up for discussion, for consideration.

There is, however, one major question which has not yet been addressed. If the establishment, or development of EMSA is seen to be even potentially desirable, one must consider how it would fit into the existing security framework of the continent. Where will it sit in relation to the multiplicity of existing international organisations in Europe with a legitimate claim to involvement in international security, even if it is a WEU subsidiary organ? In order to answer this, one must move on to consider how the overall framework of European security may evolve in the medium term - and that is the role of the concluding chapter.

ENDNOTES

[240] Cummings, D., '25 Years of British military satellite Communications', *RUSI Journal,* October 1993, p.46

[241] For a more detailed account of the origin of European civil space cooperation see McLean, A., *European Military Space Policy*, (Dartmouth, Aldershot, 1992), Chapter 4

[242] Kaminski, P., Prepared statement to the National Security sub-committee of the House Appropriations Committee, 23 March, 1995

[243] *Defense News*, Vol.6, No. 22, Jun 3-9, 1991

[244] Treaty on European Union, Article J.4.1, December 1991

[245] Javelot, M., quoted in *Defense News*, Vol.6, No. 22, 3 June, 1991

[246] Michelangelo De Maria, 'Europe in space: Edoardo Amaldi and the inception of ESRO', ESA History Project, *Document ESA HSR-5*, (ESA Publications Division, Noordwijk, Netherlands, March 1993), pp. 3-4

[247] For a detailed analysis of the legal aspect of ESA over military space policy, see Von Kries, W., 'The ESA Convention and Europe's Security Interests in Outer Space', *Zeitschrift für Luft- und Weltraumrecht, German Journal of Air and Space Law*, Vol.43, (7), 1994

CHAPTER EIGHT

A SECURITY ROLE FOR THE WEU

I. The Organisational Arena

To some observers the Western European Union (WEU) is a meaningless construct, another body which exists in the already cluttered landscape of European international organisations, many of which appear more intent on ensuring self-preservation than they are on debating the utility of their function in the post cold war environment. For others the WEU represents something quite different, the great hope of European security, an organisation with a solely European membership that can and should serve as the conduit for an independent European approach to the defence of the continent.

Although neither of these visions can lay claim to being the accepted and established truth, it can be stated with some surety that, as the United States becomes less inclined to underwrite European security in perpetuity, the issue of how Europe can best cater for its own collective defence will become of paramount importance. If the USA were to decide to leave its former partners to their own devices, the WEU and its champions would doubtless vociferously argue for it as the body to take over the mantle of the North Atlantic Treaty Organisation (NATO), replacing the perennial reliance on the transatlantic variant of security provided by that organisation with something provided by partners closer to home.

Thus the WEU finds itself at the heart of the current debate over the future shape of European security. This has manifested itself in a series of, as yet unresolved, debates regarding the status and development of the WEU: how should the WEU interact with the other international organisations it finds itself sharing an operational territory with; how will the WEU develop into an organisation capable of fulfilling an operational role in the security of Europe; and how well has the WEU adapted to the process of change which has informed the enlargement debates which so vex the EU and NATO?

II. The Security Architecture

As the twentieth century draws to a close, there remain three broad options for the WEU as it attempts to fit into the security structures required for the next century. Thus it could:

- continue to exist as a separate organisation with a solely European membership
- become the defence arm of the European Union as part of the 'deepening' process that the EU hopes to pursue
- assume the mantle of NATO - if the United States disengages itself to any significant extent from the defence of Europe

Thanks to the inertial forces associated with organisational change, the merits of both the latter options are unlikely to be tested in the near future, although if existing pronouncements are anything to go by, the most likely of these three options will be the eventual assimilation of the WEU into the structures of the European Union.

Yet, while this is something which may occur somewhere down the line, there remains a level of scepticism as to either the imminence or the likelihood of this. For example, in their report on the WEU, the British House of Commons Defence Committee include among their conclusions the following:

> 'We conclude that the IGC is unlikely to lead to any significant increase in the European Union's responsibilities in relation to defence. We do not exclude the possibility that over a much longer period WEU may eventually be subsumed in the European Union, but enlargement of the EU may complicate progress towards this.' [248]

Yet despite the possibility that the EU may come to dominate the future of European security in this way, all of the options remain subject to the vagaries of what is seen by the individual members to constitute their 'national interest' and as such are very much hostages of fortune. Clearly there are some states in Europe for whom the continued United States presence in the security of the continent rankles. Others appear keen to preserve this transatlantic linkage, while still other voices call for the preservation of sovereignty over the foreign policy process. As a result of these differences in perception, neither of the alternatives to WEU independence appears imminent at this stage. The resultant compromise position sees the WEU, NATO and the EU attempting to coexist in their overlapping constituencies. With no evidence pointing to the likelihood of impending change in this respect, it appears that the current situation is likely to remain, with all three organisations carrying on their different, if related, activities for the foreseeable future.

The significance of this fudge is that the WEU will be required to maintain its somewhat precarious existence alongside both organisations. This has undoubted benefits for the WEU, allowing as it does, at least theoretically, for independent West European military action through the provisions of the Combined Joint Task Force (CJTF) concept, and providing the equally useful side benefit of helping to separate out the defence element of European policy making from the already contentious areas that the European Union has nominal jurisdiction over. Indeed when one considers the difficulties which have been experienced in the pursuit of a single currency, the entire move towards a common European approach to security and defence through an ESDI appears optimistic. Thus while the WEU has a key role to play in the development of an ESDI, it seems likely that as pressure comes to bear on the states of Europe to subsume their individual perceptions of security and defence within a more concerted 'European' approach, the WEU will also provide a worthy forum for the discussion of such issues on a purely intergovernmental level, avoiding politically charged questions such as whether the need for unanimity will be replaced by qualified majority voting or some similar mechanism which allows for the removal of the concept of consensus.

While this attempt to maintain the current organisational framework in Europe continues, it is perhaps best viewed as an automatic 'default' setting. Thus, rather than being the result of deliberate policy choices by the member states, it represents the solution of least immediate concern. Yet viewing the WEU as a short term stop gap until some better solution is found sells the WEU short and it is by no means alone in suffering a 1990s identity crisis. If anything, the major criticism which can be levelled at the states of Western Europe over their performance in the last decade, is the fact that the real choices as regards the future shape of European security appear to have either been avoided or fudged, with the reasons for following the paths which have been chosen less than adequately explained. As an example, NATO is an organisation whose *raison d'être* evaporated with the Soviet Union, yet it has subsequently failed to convincingly assuage its critics who believe that its continued existence and desire to expand are merely the death throes of another collapsing empire. Likewise the EU has singularly failed to deliver on its own promises as contained in the Maastricht Treaty, which promised to deepen the existing union and has equally struggled to adapt to the situation in a post-Cold War Europe. Indeed expansion of the EU is even further off than it is for NATO, with the most optimistic of Central and East European states unlikely to gain access to the economic structures of Western Europe until well past 2000.

The WEU does not get off lightly in this respect either, and as a result of its relative position *vis-à-vis* NATO and the EU, has still to prove itself worthy of an active role in the security of Europe, with not all convinced that there is scope for an organisation whose claim to fame seems to be that it is a European subset of NATO.

Thus, it is possible to argue that all of the options remain very firmly on the negotiating table, with the final version of the post Cold War settlement unlikely to emerge for some time to come. That this offers the WEU an opportunity to entrench its role in this process is apparent to all, and is something that the WEU must grasp firmly if it is to retain a semblance of independence in the future. It has been the purpose of this book to assert that in some small way the Satellite Centre and more importantly the need for Europe to have a military space establishment of a more coherent nature offers the WEU among the greatest opportunities open to it in this respect. When coupled to the tacit acceptance that the WEU is here to stay and that it will be involved in military operations, either independent of NATO or in conjunction with it, as in Operation Sharp Guard, the prognosis on the organisation becomes much more upbeat.

However, despite the centrality of its role in the current organisational landscape, the WEU has been a relative newcomer to the centre stage of European security activity, until recently preferring to shun the limelight and in many cases appearing to happily cede key roles to organisations it deemed to be in a more suitable position to effectively execute them. If the WEU is to develop further it must be willing and able to fulfil an array of different roles in the preservation of European security. This is something which not all states are convinced it is capable of. The British House of Commons Defence Committee highlighted a belief that the WEU's principal role should be:

> '... to provide a structure within which European Countries can plan and execute the deployment of their military forces for humanitarian and peace-keeping tasks not involving the USA.' [249]

In the previous chapter we traced some of the more specifically space related activities that the WEU is well positioned to play a crucial role in. In this chapter we assess the broader security role the WEU can aspire to play and the importance of space based activities within this master plan. For an organisation which has spent much of its existence talking about, instead of being involved in, issues of security, this phase of adjustment and aspiration is of utmost importance.

III. The WEU and Article VIII

As has been asserted elsewhere, the need for an indigenous European intelligence gathering capability is a central component of the European move to take more control in its security destiny. The operational necessity for this is obvious, yet the independence of Europe in this respect is by no means guaranteed, despite the nascent efforts of the Satellite Centre at Torrejon.

The importance of space based assets in the future development of the WEU can not be over stated. The process of evolution which WEU has experienced since its re-emergence as a European security organisation in the mid 1980s, has allowed the WEU to legitimately consider itself a body capable of performing tasks similar to those envisaged under Chapter VIII of the United Nations Charter. Yet this is a role which does not come without costs attached, one of which is clearly the capacity to provide a comprehensive framework within which military experience gained elsewhere and the physical assets at the states' disposal can be effectively utilised.

That the WEU can consider itself a legitimate player in this field is in no small part down to the operational development which has been characterised by the concept of Combined Joint Task Forces and which will gradually be backed up by the provision of West European intelligence resources, including the data interpretation function offered by the Satellite Centre. With the emphasis placed on the role of space based assets in the execution of modern day military operations and the reliance which Europe has traditionally had on the provision of American space based resources for this purpose, the WEU's expertise in this field, and the likelihood that this will be built upon further in the future, suggests that it will soon be perfectly placed to adopt the mantle of a regional organisation capable of fulfilling a worthwhile role in the conduct of European security relations. While the most obvious candidate to play the role of security guarantor as envisaged within the UN Charter - and which Boutros Ghali's 'Agenda for Peace' further argued was necessary - remains NATO, the WEU is now a serious alternative. Indeed the Petersberg Declaration specifically states that decisions to use military units answerable to WEU will be taken by the WEU Council in accordance with the provisions of the UN Charter.[250] While it is true that the

CJTF is still a long way from being fully operational and Europe still lacks a body which can co-ordinate the military space assets which the individual states possess, the WEU can realistically be considered an actor of enormous potential in this realm.

In a strategic environment where local solutions are increasingly sought for local disputes, the WEU can be seen to have the legitimacy to fulfil this role, especially where the transatlantic linkage is deemed untenable for whatever reason. This again is an area where the WEU might be better suited to conceivable contingencies than either the European Union or NATO. [251]

IV. Avoiding Duplication

Despite the belief that the WEU can and should play a central role in the maintenance of European security, there remains a residual concern that the existence of the WEU merely replicates activities which could just as easily be carried out by either the EU or NATO. Indeed as has been discussed, it remains an option in the institutional shakedown that the WEU will eventually be subsumed within the European Union, with a vociferous caucus bemoaning the very existence of the WEU, seeing a parallel organisation as superfluous when NATO already exists.

This fear is both highlighted and challenged by Klaus Kinkel, the German Foreign Minister, who has been at pains to stress that he believes the role of the WEU need not be damaged by the existence of other security organisations. Indeed, he points to quite the opposite, feeling that there is the potential for a relationship of mutual benefit to develop between the two, although as Kinkel points out, the relationship between the two must be based upon 'complementarity and transparency' and should at all costs avoid duplication of effort.[252]

However, the potential for some form of action on a purely European level is very much on the agenda. As the former British Defence Secretary, Michael Portillo put it in a speech in 1996:

> 'The accommodation of a greater European responsibility in NATO goes to the heart of the debate about the future of European defence. For the Alliance to work well, and to remain attractive to the North American allies, Europe must demonstrate its ability to pull its weight. It is unreasonable to expect Americans to be involved in every operation of crisis management. American interest will not always be engaged. And that is where European nations must be able to take over the baton. The European military capability to do that already exists within NATO itself. Over 60 percent of the Alliance's manpower is European. This is a very large European investment in NATO's infrastructure. NATO agreed at its summit in 1994 to make Alliance assets available for European-led operations. I see the political authority and direction for European-led operations coming from the Western European Union - the WEU.' [253]

Despite such enthusiasm for the WEU, the tenor of much of the debate does point to the likely preservation of multiple organisations and the utility of having alternative frameworks designed to provide options at the regional and sub-regional levels. Whether the WEU can attain this role may depend on whether the feeling that its potential is damaged by a sense of perceived redundancy can be dispelled. As yet, the little evidence that does exist, neither supports nor disproves the WEU's ability to become an equal partner in the preservation of European security, with the WEU as yet incapable of fulfilling the roles that would enable it to offer an alternative structure to NATO.

However, there are areas where the WEU has been involved in an operational capacity - with involvements in monitoring the situation in both the Gulf conflict and its aftermath and in the more recent Yugoslav conflict.

Indeed, the latter provides an example of a situation where the WEU and NATO were involved side by side, with their joint operation in the Adriatic 'Operation Sharp Guard'. Sharp Guard offered the WEU the chance to provide naval forces to help monitor the arms embargo on the region, with the WEU providing four ships and six aircraft for the operation.[254]

While the more charitable would point to such an operation as helping provide the WEU with an operational baptism, others, such as George and

Borawski point to the 'curious fact' that the same group of states were performing the same mission but under the auspices of two separate and distinct organisations.[255]

Their interpretation of this curiosity is to find it difficult to reconcile such a joint operation with the principles set down in the Brussels Treaty, which clearly expresses the desire, as outlined in Article IV, to avoid 'the undesirability of duplicating the military staffs of NATO'.[256]

Unsurprisingly this version of events is not one subscribed to by all. To many the operations in the Adriatic and along the Danube which fell under the Adriatic Military Command were an excellent example of what can be achieved by the organisations if they can work in tandem. That the input of the WEU was more symbolic than it was expansive should not detract from the fact that it served as an introduction to the type of operation that may one day become more common. It has thus been seen as an :

> '... affirmation of the role WEU could play in future peace operations under the aegis of the United Nations or in the European framework'.[257]

The very duality of the operation can also be seen as a necessary response to a crisis which had proved difficult for any single organisation to take control of, with the eventual involvement of any organisation representing an advance on the early days of the crisis when all those which had sought to become involved, shied away from taking responsibility.

Yet while the benefits of such a joint operation are apparent, it is unlikely that the repetition of such an obvious bifurcation of operational responsibilities will be challenged by those for whom the role of multiple security organisations is hard to justify. In many ways this is a somewhat harsh reaction. NATO represents a different constituency from WEU and therefore a separate set of interests. However, as the concept of CJTF beds in, and the WEU is expected to embark on such operations on a solo basis, any repetition of this duality is far less likely to be greeted positively.

Another potential problem raised by the existence of several organisations with an interest in the security of Europe is the fear that those which do exist will shirk from their responsibility by passing the torch from one to another, with ultimately none willing or able to solve the problems of the day. Again this is a case promoted by George and Borawski who point to the fact that while states argue the case that the existence of multiple organisations is a virtue, this appears not to be evidenced by the subsequent reaction of organisations to a crisis situation. They highlight the fact that governments 'extol' the existence of the interlocking institutions which are at the heart of the European security map but then 'pass the buck among them', unable to decide which should deal with an issue at any particular time.[258]

Given that this argument was being made during the height of the Yugoslav crisis, a problem which was passed from one organisation to another like an international hot potato, this is an unsurprising sentiment and one which the peculiar circumstances of the war in Bosnia-Herzegovina helped to complicate.

Yet, for the most part it would appear that the existence of multiple organisations can and will be a valid part of the security of the twenty first century. However, that is not to say that the situation can be allowed to remain as it is. There is a clear need for the organisations to clearly delineate operational priorities and in so doing enhance the understanding of what each is expected to do. The Catch 22 situation which George and Borawski outline cannot be allowed to continue, that is to say governments must not be allowed to shirk responsibility for the resolution of crises because they are unwilling to exploit the organisations at their disposal.[259]

V. A Broader Concept of Security

With the gradual process of inclusion which has been aimed at the former Communist countries now in full swing, either through the Partnership for Peace option employed by NATO, the association agreements within the EU or the fourth tier of membership which has brought the 'Partners' into the WEU, it is obvious that the future shape of European Security has been clearly signalled. What is envisaged is a pan-European security framework which will eventually bring all the former members of the Communist bloc into the fold of the traditionally Western organisations. Thus it is being openly accepted that the security of the continent is dependent on all the states, with each and every one of them becoming involved in the preservation of peace in the continent.

However, several related questions remain and the answers to these will fundamentally affect the eventual role that the WEU plays. Thus, it is still unclear how long the Americans will remain involved in the security of continental Europe or whether they even want to, and it also remains unclear as to whether the role currently played by NATO can be performed by the Europeans alone.

The Americans themselves are adamant that the security of Europe leads to the security of the United States. This is a view clearly espoused by President Clinton as part of the National Security Strategy of the United States:

> 'European stability is vital to our security ... Our goal is an integrated democratic Europe cooperating with the United States to keep the peace and promote prosperity. The first and most important element of our strategy in Europe must be security through military strength and cooperation. The Cold War is over, but war itself is not over.' [260]

Whether this is an attitude mirrored in the corridors of power in Europe is a moot point, with some sure to question whether the Europeans actually need the United States to be involved in their security structures at all, with the French obviously at the forefront of those who would advocate the potential of Europe to look after itself without American involvement.

Indeed the Americans themselves acknowledge that the European and American perspectives are likely to diverge in the future. In an address on 'The Anglo-American Relationship', given to the Royal Institute of International Affairs in London in 1993, US Ambassador Raymond Seitz confessed:

> 'In both our countries the resources of national security are under new pressures, and our respective economic orientations are increasingly more diverse. I suspect our priorities won't match with quite the same frequency

as they once did, and the overlap of our strategic interests may not be quite so extensive as before. This is inevitable in an *ad hoc* world.' [261]

However, there are several reasons for an ongoing American participation:

- political and diplomatic weight - while Europe is respected internationally, the United States is still the most obvious power broker in the international system
- logistics - Europe still lacks the operational capacity (in all respects) of the United States and will remain dependent on the Americans for some time to come
- desire - both the public and the administration remain convinced of the need for ongoing United States involvement in the preservation of security in Europe

Thus the Partnership for Peace and the interwoven pattern of organisational enlargement represents a major element in the provision of stability in Europe and the continuance of a linkage between the United States and its European partners. However, if the process of NATO enlargement is at the heart of European security for the start of the next millennium the question is begged, why do we also need an organisation representing a European subset of the premier security body.

Again there are several arguments which can be made and need not be deemed as contradictory to those made in favour of the preservation of the transatlantic link.

- The United States does not want to be permanently involved in European security affairs at all levels and in all circumstances (indeed neither would Europe wish the USA to be involved in all of these). As a result there is a clear rationale for some kind of European self sufficiency in this area and an organisation tasked to service it, thus providing an alternative framework and also attempting to avoid a damaging and debilitating repeat of the burden sharing dilemma of the 1980s and before.

- Europe seems unlikely to develop a fully integrated security and defence identity which will allow for the conduct of a truly European foreign policy. The natural vehicle for such a foreign policy is the European Union, yet in lieu of such an eventuality some kind of consensus based inter-governmental structure will be vital. In this respect the WEU is developing its ability to perform this function and is perfectly placed to promote itself in this role.
- As a consequence of being a hybrid organisation, in that it couples the inter-governmental ethos while retaining a more narrow membership than an organisation such as the OSCE, the WEU is the one body where the militarily powerful states in Europe maintain a major influence, while not being able to ride roughshod over the other members.

This third point is of importance because it demonstrates that in the case of a crisis to be dealt with by Europe and without anything other than diplomatic support from the United States - a circumstance in which Europe would inevitably be looking to the British, the French and the Germans to take the political and military initiative - they would have the scope to use their leverage while being constrained by their traditional allies, not the entire European supporting cast. This suggests a clear demarcation between ranks of European states, something which the three tier European security structure as envisaged by Jemiolo hints at.[262] This model sees an A list of states which includes the current (and future) membership of NATO and the WEU which are deemed secure states. These are followed by the B list of non-NATO countries which are part of the 'partnership' activities and fall under the umbrella of the NACC. This group can also be sub-divided to include those which are likely to graduate into the A list, with the likes of Poland, Hungary and the Czech Republic states which by dint of geography, history and politics find themselves a part of the traditional Western European security landscape and at the front of the queue to join in a first phase of enlargement. There are others who find themselves in a less preferential position vis-à-vis the likes of these three Visegrad states, with Bulgaria and Slovakia as well as the Baltic Republics the most obvious examples of states which are neither fully secure nor entirely ignored. The third grouping is the assorted group of remaining states, those who are for whatever reason disenfranchised from the security frameworks of Europe.

If we apply a very simple economic model to this inter-relationship, we can further sub-divide Jemiolo's first grouping and suggest the existence of a consumer-producer dynamic, with the aforementioned British, French and Germans replacing the USA as the providers of security in a future European model.

VI. The Expansion Process

The need to co-ordinate the enlargement policies of the various organisations is obvious. While the actual dates of entry into the NATO or the EU may be different, there will clearly be a symbiosis between the two processes. Thus, it is to be expected that there will be overlapping memberships. Serfaty points to the anomaly that this clearly creates, with the current role of Turkey very much one of interest.[263] As a member of the NATO fold but not yet accepted as a WEU member, the Turks are deemed to be at the periphery of West European affairs, despite being considered a crucial player in NATO. This situation is clearly replicated in the EU, where the wariness of Western Europe towards Turkish entry into the mainstream of European economic and political activities is still resisted. The vexed question of the status of Turkey is thus another issue which requires attention.

VII. The WEU at Fifty

The Brussels Treaty reaches the ripe old age of fifty in 1998. It is suggested that this juncture will see the Treaty and its contents come up for re-negotiation, with the entire future of the organisation possibly coming under consideration. However, this is based on a misunderstanding of the text of the Treaty. The significance of the fifty year period lies in Article 10 of the original Treaty (renumbered as Article 12 following the modifications made in Paris) which does indeed state that the Treaty shall remain in force for fifty years. However, the same article goes on to state:

> 'After the expiry of the period of fifty years, each of the High Contracting Parties shall have the right to cease to be a party thereto provided that he shall have given one year's notice of denunciation to the Belgian Government.'

A closer reading of the Treaty thus clearly demonstrates that, after the fifty year period, the Treaty will continue in force. The only difference being that after 1998, any of the parties may cease to be a party thereto. There is nothing, however, within the Treaty itself to suggest that it ceases to exist after the expiry of fifty years and thus there is no need for a re-negotiation of its terms and commitments.

What is much more likely is that the 50th anniversary will lead to a much greater awareness of the organisation and attention may well be diverted to the role that the WEU, both at present and in the future, will be expected to play. Thus, the expiry of fifty years may prove to be a happy coincidence, with the opportunity it affords to discuss the activities of the WEU coming at the perfect time for the organisation to seize the initiative as regards European military space and other operational requirements.

One of the crucial features of this discussion may well be the level to which the WEU can perform a significantly independent role within the European security landscape, with a full scale evaluation of where the WEU sits in relation to the other security organisations and how it can hope to fulfil an independent role when its tasks are almost inevitably duplicated to some extent by a combination of other organisations.

Yet despite these apparent problems, most observers concur that the stock of the WEU is on the rise and has been since the Treaty on European Union was signed and the role of the WEU in the development of the security and defence identity in Europe was formally recorded.

In 1994, the then Secretary General of the WEU, Dr. Willem van Eekelen, described the role as:

> 'Being at the same time the defence component of the Union, the potential tool of the emerging Common Foreign and Security Policy and the European pillar of the renewed transatlantic partnership on which our ultimate security continues to depend, Western European Union has a role rather like a hinge. It serves as a hinge between two of the fundamental requirements of our common security on the European continent. The first is to maintain the link with our North American friends and Allies in the defence field. The second is to develop the European Union, with all the tools of preventive diplomacy it needs to contribute to project peace and stability across Europe for the benefit of all Europeans.' [264]

Of the various roles that are being actively considered for it, the WEU does fit comfortably into the mould of an organisation which could deal with the militarisation of West European space policy. With the introduction of the Satellite Centre and the widespread belief that the Europeans will have to become ever more heavily involved in the exploitation of space based resources if they are to successfully carry out missions without at least the tacit support of the United States, then the ongoing development of space for military activities will have to be a characteristic of the priorities which military planners adopt. While this is already happening on a rather more ad hoc basis, with the individual and collaborative use of space based assets, there is clearly a rationale for the concerted efforts of the European partners via some forum, to come together and push this agenda forward. While there already exists a European Space Agency in the form of ESA, this is designed to satisfy the civilian market. To suggest that ESA could move comfortably into the co-ordination of military space activities is to stretch credulity. Yet the WEU has over many decades now demonstrated its interest in this area and makes the ideal example of a body which is capable and willing to perform this task.

What remains is for the politicians and the national governments to recognise the need to address this issue and have the foresight to fund the development of European resources that, while being unlikely to match those available to the USA, should allow Europe to hold its own in the majority of the areas outlined in Chapter Six.

There seems to be a considerable level of consensus amongst European states as to the important role which the WEU can and indeed must play in European security.

The French for example are among the most vociferous backers of the WEU. This comes across most notably in French support for the notion of the WEU developing its role as an independent actor in security activities should NATO be either unable or unwilling to become involved in certain matters deemed 'European' problems. To this end the French return to the military fold of NATO can perhaps be viewed as an attempt to extend its political influence within the military arm of NATO and aid its lobbying weight on behalf of the WEU. That the operational capability of the WEU is currently being clarified under the aegis of the 'Combined Joint Task Force' (CJTF) initiative is in itself further evidence of the potential activities that the WEU could find itself becoming involved in.

Even the British, not normally regarded as the most willing participants in integrated European activities, have espoused the view that the WEU is highly important. Addressing the IGC Reflection Group Seminar in January 1996, David Davis, minister of state at the Foreign and Commonwealth Office said:

> 'NATO remains the only solution for our own direct defence. But there will be times when a peacekeeping, humanitarian or other crisis-management mission is called for and when the US and Canada for some good reason will not be involved. We believe the WEU is the best and only organisation to muster and command European forces for these ends.' [265]

VIII. Conclusions

This book has traced the progress of the WEU as both an organisation in the broader sense and as a key component in the discussion and examination of the role that Western Europe should be playing in the exploitation of space based assets. This has seen us trace the WEU from the optimism which surrounded the creation of the organisation in 1954, to the brink of the abyss during the 1970s and on from that low point into the brighter future which the 1990s and beyond appears to hold for an organisation which has spent much of its existence in suspended animation.

If Europe is serious about the pursuit of a defence and security identity which can realistically attempt to offer an alternative to the organisational framework provided by NATO, then the WEU is the only viable body that can take on the role. Similarly, if a significant degree of European autonomy in the field of security is a valid aspiration, then Europe must also aspire to a coherent military space policy.

While it could be argued that the very survival of the WEU through to the 1990s is evidence of a degree of success beyond prediction two decades ago, it is the development of the organisation in the next few years that will be of vital importance in any long term prognosis of its health. At the heart of the WEU's ability to serve an independent function is the Satellite Centre and it remains the contention of this book that the role of the Satellite Centre and space based systems in general, are not only important elements in the survival of the WEU, but are vital to the security of Europe.

ENDNOTES

[248] House of Commons Defence Committee, Fourth Report, Session 1995-96, *Western European Union*, (London, HMSO, 1996) p xi

[249] House of Commons Defence Committee, Ibid., p xiii

[250] See WEU Petersberg Declaration, included as Appendix Five

[251] This idea was forwarded by Baarveld-Schlaman, *Information letter from the Assembly of the WEU*, No.16, July 1993

[252] Kinkel, K., 'The new NATO: Steps towards reform', *NATO Review* No.3, May 1996, p11

[253] Portillo, M., Speech by the Secretary State for Defence on *International Security*, delivered in Washington, 23 January 1996

[254] WEU Secretariat-General, *WEU Today*, (Brussels, Belgium, 1997) p26

[255] George, B., & Borawski, J., 'Sympathy for the Devil: European Security in a Revolutionary Age', *European Security* Vol.2, No.4, Winter 1993, p480

[256] George, B., & Borawski, J., 'Sympathy for the Devil: European Security in a Revolutionary Age', *European Security* Vol.2, No.4, Winter 1993, p480

[257] These points were raised by Beniamino Andreatta, Italian Minister of Foreign Affairs, at the 39th Session of the Assembly of the WEU, *Information letter from the Assembly of the WEU*, No.16, July 1993

[258] George, B., & Borawski, J., 'Sympathy for the Devil: European Security in a Revolutionary Age', *European Security*, Vol.2, No.4, Winter 1993, p476

[259] George, B., & Borawski, J., Ibid., p476

[260] The White House, *A National Security Strategy of Engagement and Enlargement*, (Washington, US Govt. Printing Office, Feb 1996)

[261] US Ambassador Seitz, R.G.H., *The Anglo-American Relationship*, a speech given to the Royal Institute of International Affairs, Chatham House, London, 23 March 1993

[262] Jemiolo, T., 'Fundamental Issues of Security in Europe', *European Security*, Vol.2, No.3, Autumn 1993, p381

[263] Serfaty, S., 'America and Europe Beyond Bosnia', *The Washington Quarterly*, Vol.19:3, 1996, p.43

[264] Secretary General of the WEU, van Eekelen, W., *WEU after two Brussels Summits: A new approach to common tasks*, presentation to the Royal Institute for International Relations, Brussels, 27 January 1994

[265] Davis, D., Speech by the FCO Minister *on WEU and EU: Britain's Ambitions for European Defence*, delivered in Oslo, 22 January 1996 to the IGC Reflection Group Seminar

APPENDIX ONE

WESTERN EUROPEAN UNION
BRUSSELS TREATY

As amended by the Protocol modifying and completing the Brussels Treaty, signed at Paris on October 23, 1954

TREATY OF ECONOMIC, SOCIAL AND CULTURAL COLLABORATION AND COLLECTIVE SELF-DEFENCE SIGNED AT BRUSSELS ON MARCH 17, 1948, AS AMENDED BY THE "PROTOCOL MODIFYING AND COMPLETING THE BRUSSELS TREATY"

[The High Contracting Parties]

Resolved:

To reaffirm their faith in fundamental human rights, in the dignity and worth of the human person and in the other ideals proclaimed in the Charter of the United Nations;

To fortify and preserve the principles of democracy, personal freedom and political liberty, the constitutional traditions and the rule of law, which are their common heritage;

To strengthen, with these aims in view, the economic, social and cultural ties by which they are already united;

To co-operate loyally and to co-ordinate their efforts to create in Western Europe a firm basis for European economic recovery;

To afford assistance to each other, in accordance with the Charter of the United Nations, in maintaining international peace and security and in resisting any policy of aggression;

To promote the unity and to encourage the progressive integration of Europe;

To associate progressively in the pursuance of these aims other States inspired by the same ideals and animated by the like determination;

Desiring for these purposes to conclude a treaty of collaboration in economic, social and cultural matters and for collective self-defence;

Have agreed as follows:

ARTICLE I

Convinced of the close community of their interests and of the necessity of uniting in order to promote the economic recovery of Europe, the High Contracting Parties will so organise and co-ordinate their economic activities as to produce the best possible results, by the elimination of conflict in their economic policies, the co-ordination of production and the development of commercial exchanges.

The co-operation provided for in the preceding paragraph, which will be effected through the Council referred to in Article VIII, as well as through other bodies, shall not involve any duplication of, or prejudice to, the work of other economic organisations in which the High Contracting Parties are or may be represented, but shall on the contrary assist the work of those organisations.

ARTICLE II

The High Contracting Parties will make every effort in common, both by direct consultation and in specialised agencies, to promote the attainment of a higher standard of living by their peoples and to develop on corresponding lines the social and other related services of their countries.

The High Contracting Parties will consult with the object of achieving the earliest possible application of recommendations of immediate practical interest, relating to social matters, adopted with their approval in the specialised agencies.

They will endeavour to conclude as soon as possible conventions with each other in the sphere of social security.

ARTICLE III

The High Contracting Parties will make every effort in common to lead their peoples towards a better understanding of the principles which form the basis of their common civilisation and to promote cultural exchanges by conventions between themselves or by other means.

ARTICLE IV

In the execution of the Treaty, the High Contracting Parties and any Organs established by Them under the Treaty shall work in close co-operation with the North Atlantic Treaty Organisation.

Recognising the undesirability of duplicating the military staffs of NATO, the Council and its Agency will rely on the appropriate military authorities of NATO for information and advice on military matters.

ARTICLE V

If any of the High Contracting Parties should be the object of an armed attack in Europe, the other High Contracting Parties will, in accordance with the provisions of Article 51 of the Charter of the United Nations, afford the Party so attacked all the military and other aid and assistance in their power.

ARTICLE VI

All measures taken as a result of the preceding Article shall be immediately reported to the Security Council. They shall be terminated as soon as the Security Council has taken the measures necessary to maintain or restore international peace and security.

The present Treaty does not prejudice in any way the obligations of the High Contracting Parties under the provisions of the Charter of the United Nations. It shall not be interpreted as affecting in any way the authority and responsibility of the Security Council under the Charter to take at any time such action as it deems necessary in order to maintain or restore international peace and security.

ARTICLE VII

The High Contracting Parties declare, each so far as he is concerned, that none of the international engagements now in force between him and any other of the High Contracting Parties or any third State is in conflict with the provisions of the present Treaty.

None of the High Contracting Parties will conclude any alliance or participate in any coalition directed against any other of the High Contracting Parties.

ARTICLE VIII

1. For the purposes of strengthening peace and security and of promoting unity and of encouraging the progressive integration of Europe and closer co-operation between Them and with other European organisations, the High Contracting Parties to the Brussels Treaty shall create a Council to consider matters concerning the execution of this Treaty and of its Protocols and their Annexes.
2. This Council shall be known as the "Council of Western European Union"; it shall be so organised as to be able to exercise its

functions continuously; it shall set up such subsidiary bodies as may be considered necessary: in particular it shall establish immediately an Agency for the Control of Armaments whose functions are defined in Protocol No. IV.

3. At the request of any of the High Contracting Parties the Council shall be immediately convened in order to permit Them to consult with regard to any situation which may constitute a threat to peace, in whatever area this threat should arise, or a danger to economic stability.
4. The Council shall decide by unanimous vote questions for which no other voting procedure has been or may be agreed. In the cases provided for in Protocols II, III and IV it will follow the various voting procedures, unanimity, two-thirds majority, simple majority, laid down therein. It will decide by simple majority questions submitted to it by the Agency for the Control of Armaments.

ARTICLE IX

The Council of Western European Union shall make an annual report on its activities and in particular concerning the control of armaments to an Assembly composed of representatives of the Brussels Treaty Powers to the Consultative Assembly of the Council of Europe.

ARTICLE X

In pursuance of their determination to settle disputes only by peaceful means, the High Contracting Parties will apply to disputes between themselves the following provisions;

The High Contracting Parties will, while the present Treaty remains in force, settle all disputes falling within the scope of Article 36, paragraph 2, of the Statute of the International Court of Justice, by referring them to the Court, subject only, in the case of each of them, to any reservation already made by that Party when accepting this clause for compulsory jurisdiction to the extent that that Party may maintain the reservation. In addition, the High Contracting Parties will submit to conciliation all disputes outside the scope of Article 36, paragraph 2, of the Statute of the International Court of Justice.

In the case of a mixed dispute involving both questions for which conciliation is appropriate and other questions for which judicial settlement is appropriate, any Party to the dispute shall have the right to insist that the judicial settlement of the legal questions shall precede conciliation.

The preceding provisions of this Article in no way affect the application of relevant provisions or agreements prescribing some other method of pacific settlement.

ARTICLE XI

The High Contracting Parties may, by agreement, invite any other State to accede to the present Treaty on conditions to be agreed between them and the State so invited.

Any State so invited may become a Party to the Treaty by depositing an instrument of accession with the Belgian Government.

The Belgian Government will inform each of the High Contracting Parties of the deposit of each instrument of accession.

ARTICLE XII

The present Treaty shall be ratified and the instruments of ratification shall be deposited as soon as possible with the Belgian Government.

It shall enter into force on the date of the deposit of the last instrument of ratification and shall thereafter remain in force for fifty years.

After the expiry of the period of fifty years, each of the High Contracting Parties shall have the right to cease to be a party thereto provided that he shall have previously given one year's notice of denunciation to the Belgian Government.

The Belgian Government shall inform the Governments of the other High Contracting Parties of the deposit of each instrument of ratification and of each notice of denunciation.

Appendix Two

North Atlantic Treaty Organisation
Washington Treaty

The North Atlantic Treaty signed at Washington D.C. on April 4, 1949

The Parties to this Treaty reaffirm their faith in the purposes and principles of the Charter of the United Nations and their desire to live in peace with all peoples and all governments.

They are determined to safeguard the freedom, common heritage and civilisation of their peoples, founded on the principles of democracy, individual liberty and the rule of law.

They seek to promote stability and well-being in the North Atlantic area.

They are resolved to unite their efforts for collective defence and for the preservation of peace and security.

They therefore agree to this North Atlantic Treaty:

Article 1

The Parties undertake, as set forth in the Charter of the United Nations, to settle any international disputes in which they may be involved by peaceful means in such a manner that international peace and security and justice are not endangered, and to refrain in their international relations from the threat or use of force in any manner inconsistent with the purposes of the United Nations.

Article 2

The Parties will contribute toward the further development of peaceful and friendly international relations by strengthening their free institutions, by bringing about a better understanding of the principles upon which these institutions are founded, and by promoting conditions of stability and well-being. They will seek to eliminate conflict in their international economic policies and will encourage economic collaboration between any or all of them.

Article 3

In order more effectively to achieve the objectives of this Treaty, the Parties, separately and jointly, by means of continuous and effective self-help and mutual aid, will maintain and develop their individual and collective capacity to resist armed attack.

Article 4

The Parties will consult together whenever, in the opinion of any of them, the territorial integrity, political independence or security of any of the Parties is threatened.

Article 5

The Parties agree that an armed attack against one or more of them in Europe or North America shall be considered an attack against them all; and consequently they agree that, if such an armed attack occurs, each of them, in exercise of the right of individual or collective self-defence recognised by Article 51 of the Charter of the United Nations, will assist the Party or Parties so attacked by taking forthwith, individually and in concert with the other

Parties, such action as it deems necessary, including the use of armed force, to restore and maintain the security of the North Atlantic area.

Any such armed attack and all measures taken as a result thereof shall immediately be reported to the Security Council. Such measures shall be terminated when the Security Council has taken the measures necessary to restore and maintain international peace and security.

Article 6

For the purpose of Article 5, an armed attack on one or more of the Parties is deemed to include an armed attack:

-- on the territory of any of the Parties in Europe or North America, on the Algerian Departments of France, on the territory of Turkey or on the Islands under the jurisdiction of any of the Parties in the North Atlantic area north of the Tropic of Cancer;

-- on the forces, vessels, or aircraft of any of the Parties, when in or over these territories or any other area in Europe in which occupation forces of any of the Parties were stationed on the date when the Treaty entered into force or the Mediterranean Sea or the North Atlantic area north of the Tropic of Cancer.

Article 7

This Treaty does not affect, and shall not be interpreted as affecting, in any way the rights and obligations under the Charter of the Parties which are members of the United Nations, or the primary responsibility of the Security Council for the maintenance of international peace and security.

Article 8

Each Party declares that none of the international engagements now in force between it and any other of the Parties or any third State is in conflict with the provisions of this Treaty, and undertakes not to enter into any international engagement in conflict with this Treaty.

Article 9

The Parties hereby establish a Council, on which each of them shall be represented, to consider matters concerning the implementation of this Treaty. The Council shall be so organised as to be able to meet promptly at any time. The Council shall set up such subsidiary bodies as may be necessary; in particular it shall establish immediately a defence committee which shall recommend measures for the implementation of Articles 3 and 5.

Article 10

The Parties may, by unanimous agreement, invite any other European state in a position to further the principles of this Treaty and to contribute to the security of the North Atlantic area to accede to this Treaty. Any state so invited may become a Party to the Treaty by depositing its instrument of accession with the Government of the United States of America. The Government of the United States of America will inform each of the Parties of the deposit of each such instrument of accession.

Article 11

This Treaty shall be ratified and its provisions carried out by the Parties in accordance with their respective constitutional processes. The instruments of ratification shall be deposited as soon as possible with the Government of the United States of America, which will notify all the other signatories of each deposit. The Treaty shall enter into force between the States which have ratified it as soon as the ratifications of the majority of the signatories, including the ratifications of Belgium, Canada, France, Luxembourg, the Netherlands, the United Kingdom and the United States, have been deposited and shall come into effect with respect to other States on the date of the deposit of their ratifications.

Article 12

After the Treaty has been in force for ten years, or at any time thereafter, the Parties shall, if any of them so requests, consult together for the purpose of reviewing the Treaty, having regard for the factors then affecting peace and security in the North Atlantic area, including the development of universal as well as regional arrangements under the Charter of the United Nations for the maintenance of international peace and security.

Article 13

After the Treaty has been in force for twenty years, any Party may cease to be a Party one year after its notice of denunciation has been given to the Government of the United States of America, which will inform the Governments of the other Parties of the deposit of each notice of denunciation.

Article 14

This Treaty, of which the English and French texts are equally authentic, shall be deposited in the archives of the Government of the United States of America. Duly certified copies will be transmitted by that Government to the Governments of the other signatories.

APPENDIX THREE

WESTERN EUROPEAN UNION VIANDEN MINISTERIAL

Ministerial decision setting up a WEU Satellite Centre
taken at Vianden, Luxembourg, June 27, 1991

The Ministerial Council:

- Recalling its decision of 10 December 1990 on space cooperation within WEU,
- Approving the results of the work of the ad hoc Sub Group on Space regarding the setting-up of a WEU Satellite Centre,

Decides as follows:

1. A WEU subsidiary body will be set up, in accordance with Article VIII, paragraph 2, of the modified Brussels Treaty, to be known as the WEU Satellite Centre, hereinafter referred to as the Centre.
2. The Centre will be placed under the authority of the WEU Council. Its legal personality will not be separate from that of the WEU. The Centre's managing bodies will be the Director and the Steering Committee.

3. The Director will be appointed for a term of three years by the WEU Council. He will be responsible for the management of the Centre. He will exercise hierarchical authority over the Centre's staff.
4. Each member State will appoint a representative to the Centre's Steering Committee. The latter will lay down technical and financial guidelines to be implemented by the Centre, and be in constant liaison with the national authorities calling upon the Centre's services.
5. The Centre will offer no legal guarantee as to services supplied in accordance with this decision, and WEU will incur no liability to member States by virtue of this fact.
6. The financial resources of the Centre will be provided for in a separate budget approved annually by the WEU Permanent Council in accordance with the current financial regulations of WEU. The budget shall be financed by contributions from member States in accordance with a cost-sharing formula agreed by the Council. The Centre's budget for the experimental phase will be within a maximum ceiling of 38.25 million ECUs, based on an estimate of 30.6 million ECUs plus 25% for contingencies.
7. The Centre will be established in the territory of one of the WEU's member States.
8. The personnel responsible for setting up the Centre will be appointed by member States or on the recommendation of the ad hoc Group before September 1991. The Centre will be established with effect from 1 January 1992 and will enter into operations later that year.
9. At their next meeting, a report reviewing progress on the establishment of the Centre. and a draft framework agreement covering access by the Centre to Helios data, will be presented to Ministers, who will decide on the location of the Centre and the appointment of a Director.
10. The activities of the Centre should be reviewed no later than three years after it has entered into operation, when it will be possible for each member State to reconsider its participation in the Centre.

The possibilities for future cooperation between the Centre and other corresponding bodies, particularly in Europe and in the United States, on the basis of a balanced partnership.

APPENDIX FOUR

EUROPEAN UNION DECLARATION ON WESTERN EUROPEAN UNION

Declaration on Western European Union taken from the Final Act of the Treaty on European Union signed at Maastricht, December 1991

I. DECLARATION

by Belgium, Germany, Spain, France, Italy, Luxembourg, the Netherlands, Portugal and the United Kingdom of Great Britain and Northern Ireland, which are members of the Western European Union and also members of the European Union on

THE ROLE OF THE WESTERN EUROPEAN UNION AND ITS RELATIONS WITH THE EUROPEAN UNION AND WITH THE ATLANTIC ALLIANCE

Introduction

1. WEU Member States agree on the need to develop a genuine European security and defence identity and a greater European responsibility on defence matters. This identity will be pursued through a gradual process involving successive phases. WEU will

form an integral part of the process of the development of the European Union and will enhance its contribution to solidarity within the Atlantic Alliance. WEU Member States agree to strengthen the role of WEU, in the longer term perspective of a common defence, compatible with that of the Atlantic Alliance.

2. WEU will be developed as the defence component of the European Union and as a means to strengthen the European pillar of the Atlantic Alliance. To this end, it will formulate common European defence policy and carry forward its concrete implementation through the further development of its own operational role. WEU Member States take note of Article J.4 relating to the common foreign and security policy of the Treaty on European Union which reads as follows:

   1. The common foreign and security policy shall include all questions related to the security of the Union, including the eventual framing of a common defence policy, which might in time lead to a common defence.
   2. The union requests the Western Union (WEU), which is an integral part of the development of the Union, to elaborate and implement decisions and actions of the Union which have defence implications. The Council shall, in agreement with the institutions of the WEU, adopt the necessary practical arrangements.
   3. Issues having defence implications dealt with under this Article shall not be subject to the procedures set out in Article J.3.
   4. The policy of the Union in accordance with this Article shall not prejudice the specific character of the security and defence policy of certain Member States and shall respect the obligations of certain Member States under the North Atlantic Treaty and be compatible with the common security and defence policy established within that framework.

5. The provisions of this Article shall not prevent the development of closer cooperation between two or more Member States on a bilateral level, in the framework of the WEU and the Atlantic Alliance, provided such cooperation does not run counter to or impede that provided for in this Title.
6. With a view to furthering the objective of this Treaty, and having in view the date of 1998 in the context of Article XII of the Brussels Treaty, the provisions of this Article may be revised as provided for in Article N(2) on the basis of a report to be presented in 1996 by the Council to the European Council, which shall include an evaluation of the progress made and the experience gained until then.

A. WEU's relations with European Union

3. The objective is to build up WEU in stages as the defence component of the European Union. To this end, WEU is prepared, at the request of the European Union, to elaborate and implement decisions and actions of the Union which have defence implications. To this end, WEU will take the following measures to develop a close working relationship with the Union:

- as appropriate, synchronization of the dates and venues of meetings and harmonization of working methods;
- establishment of close cooperation between the Council and Secretariat- General of WEU on the one hand, and the Council of the Union and General Secretariat of the Council on the other;
- consideration of the harmonization of the sequence and duration of the respective Presidencies;
- arranging for appropriate modalities so as to ensure that the Commission of the European Communities is regularly

informed and, as appropriate, consulted on WEU activities in accordance with the role of the Commission in the common foreign and security policy as defined in the Treaty on European Union;

- encouragement of closer cooperation between the Parliamentary Assembly of WEU and the European Parliament.

The WEU Council shall, in agreement with the competent bodies of the European Union, adopt the necessary practical arrangements.

B. WEU's relations with the Atlantic Alliance

4. The objective is to develop WEU as a means to strengthen the European pillar of the Atlantic Alliance. Accordingly WEU is prepared to develop further the close working links between WEU and the Alliance and to strengthen the role, responsibilities and contributions of WEU Member States in the Alliance. This will be undertaken on the basis of the necessary transparency and complementarity between the emerging European security and defence identity and the Alliance. WEU will act in conformity with the positions adopted in the Atlantic Alliance.

- WEU Member States will intensify their coordination on Alliance issues which represent an important common interest with the aim of introducing joint positions agreed in WEU into the process of consultation in the Alliance which will remain the essential forum for consultation among its members and the venue for agreement on policies bearing on the security and defence commitments of Allies under the North Atlantic Treaty.
- Where necessary, dates and venues of meetings will be synchronized and working methods harmonized.

- Close cooperation will be established between the Secretariats-General of WEU and NATO.

C. Operational role of WEU

5. WEU's operational role will be strengthened by examining and defining appropriate missions, structures and means, covering in particular:

- WEU planning cell;
- closer military cooperation complementary to the Alliance in particular in the fields of logistics, transport, training and strategic surveillance;
- meetings of WEU Chiefs of Defence Staff;
- military units answerable to WEU.

Other proposals will be examined further including:

- - enhanced cooperation in the field of armaments with the aim of creating a European armaments agency;
- - development of the WEU Institute into a European Security and Defence Academy.

Arrangements aimed at giving WEU a stronger operational role will be fully compatible with the military dispositions necessary to ensure the collective defence of all Allies.

D. Other measures

1. As a consequence of the measures set out above, and in order to facilitate the strengthening of WEU's role, the seat of the WEU Council and Secretariat will be transferred to Brussels.
2. Representation of the WEU Council must be such that the Council is able to exercise its functions continuously in accordance with Article VIII of the modified Brussels Treaty. Member States may draw on a double-hatting formula, to be worked out, consisting of their representatives to the Alliance and to the European Union.
3. WEU notes that, in accordance with the provisions of Article J.4(6) concerning the common foreign and security policy of the Treaty on European Union, the Union will decide to review the provisions of this Article with a view to furthering the objective to be set by it in accordance with the procedure defined. The WEU will re-examine the present provisions in 1996. This re-examination will take account of the progress and experience acquired and will extend to relations between WEU and the Atlantic Alliance.

II. DECLARATION

by Belgium, Germany, Spain, France, Italy, Luxembourg, the Netherlands, Portugal and the United Kingdom of Great Britain and Northern Ireland which are members of the Western European Union.

"The Member States of WEU welcome the development of the European security and defence identity. They are determined, taking into account the role of WEU as the defence component of the European Union and as the means to strengthen the European pillar of the Atlantic Alliance, to put the relationship between WEU and the other European States on a new basis for the sake of stability and security in Europe. In this spirit, they propose the following: States which are members of the European Union are invited to accede to WEU on conditions to be agreed in accordance with Article XI of the modified Brussels Treaty, or to become observers if they so wish. Simultaneously, other European Member States of NATO are invited to become associate members of WEU in a way which will give them the possibility of participating fully in the activities of WEU.

The Member States of WEU assume that treaties and agreements corresponding with the above proposals will be concluded before 31 December 1992."

APPENDIX FIVE

WESTERN EUROPEAN UNION PETERSBERG DECLARATION

Western European Union, Council of Ministers,
Petersberg Declaration signed at Bonn, 19 June 1992

The Foreign and Defence Ministers of WEU member States met in Bonn on 19 June 1992 and issued the Petersberg Declaration consisting of the following three parts:

I. On WEU and European Security
II. On strengthening WEU's operational role
III. On relations between WEU and the other European member States of the European Union or the Atlantic Alliance

I. ON WEU AND EUROPEAN SECURITY

Developments in the security situation in Europe, disarmament and arms control

1. Ministers reviewed the significant changes that had taken place in the security situation in Europe since their last regular meeting in November 1991. They emphasized the importance of strengthening the role and institutions of the CSCE for peace and security in Europe. They looked

forward to decisions at Helsinki to start new negotiations on measures of arms control and disarmament and to enhance regular consultations and cooperation on security matters. In the light of the establishment of a new CSCE Forum for Security Cooperation, they considered that decisions to enhance the CSCE's capabilities for conflict prevention, crisis management and the peaceful settlement of disputes are of primary importance. They supported the proposal under discussion at the Helsinki Follow-up meeting for the CSCE to declare itself as a regional arrangement under Chapter VIII of the United Nations Charter. Ministers considered that the CSCE should have the authority to initiate and pursue peacekeeping operations under its own responsibility.

2. As WEU develops its operational capabilities in accordance with the Maastricht Declaration, we are prepared to support, on a case-by-case basis and in accordance with our own procedures, the effective implementation of conflict-prevention and crisis-management measures, including peacekeeping activities of the CSCE or the United Nations Security Council. This will be done without prejudice to possible contributions by other CSCE countries and other organisations to these activities.

3. Ministers welcomed the decisions taken by the CSCE Council in Berlin and Prague regarding the relationships between the CSCE and other mutually reinforcing European and transatlantic organizations including WEU. They declared that WEU, together with the European Union, was ready to play a full part in building up Europe's security architecture. They likewise reaffirmed their conviction that the Atlantic Alliance is one of the indispensable foundations of Europe's security. They welcomed the ongoing reform process of NATO with a view to establishing a strong new transatlantic partnership.

4. Ministers welcomed the agreement reached at the CFE Extraordinary Conference on 5 June 1992 in Oslo which provides the basis for the entry into force of the CFE Treaty which has been and remains a major

objective of their arms control agenda. Its full and effective implementation will increase stability and open the way to a new cooperative security order in Europe. They call upon the new States parties to the Treaty to ensure its ratification by the time of the CSCE Summit in Helsinki. Ministers attach great importance to the conclusion of an agreement on the limitation of personnel strengths of ground and airforces (CFE 1a) in time for the Helsinki Summit and to the implementation of the Open Skies Treaty. They reaffirmed their commitment to the early entry into force of the Open Skies agreement and invited other CSCE States to accede to the Treaty in accordance with its provisions.

5. Ministers welcomed steps recently taken by the States concerned to allow for the entry into force of the START Treaty and the important agreement on further strategic reductions reached between the United States and Russia in Washington on 17 June 1992.

6. Ministers recalled that the presence of foreign forces on the territory of a sovereign state requires the explicit consent of that state. They stressed the importance of rapidly establishing, in the negotiations under way, timetables for the withdrawal of foreign troops from the territory of the Baltic States.

7. Ministers expressed their conviction that a Chemical Weapons Convention can be reached within the next few months. They are confident that this Convention can play an important and pioneering role in worldwide multilateral arms control and call on all member States of the Conference on Disarmament to lend their support to the emerging consensus. They repeat their commitment to be among the original signatories of this Convention and ask all other nations to follow this course.

8. WEU member States reaffirmed their resolution to contribute further to the establishment of a new order of peace in Europe which, in accordance

with the Charter of Paris, will be based on cooperation. Ministers underlined the valuable contribution of NACC in this connection. In the same spirit, WEU has invited the Foreign and Defence Ministers of eight States of Central Europe to a special Ministerial meeting later today. WEU and the invited countries intend to enhance consultation and cooperation in the framework of the new European security structure.

Implementation of the Maastricht Declaration

9. Ministers stressed the fundamental importance of the Treaty on European Union and they looked forward to the further elaboration of the common foreign and security policy at the Lisbon European Council. They discussed the progress made in developing the role of WEU as the defence component of the European Union and as the means to strengthen the European pillar of the Atlantic Alliance in accordance with the Declaration adopted by WEU member States at the Maastricht European Council in December 1991.

10. Ministers reaffirmed the importance for WEU to develop close working relations with the European Union and Atlantic Alliance in accordance with the Maastricht Declaration of WEU. They adopted a report on the practical measures necessary for WEU to develop these relations. They asked the Permanent Council to propose to the Council of the Twelve and to the North Atlantic Council concrete measures aimed at facilitating the development of close cooperation between the respective Secretariats.

11. Ministers heard a report from the Secretary-General on the progress made towards the transfer of the WEU Council and Secretariat-General from London to Brussels. They instructed the Permanent Council and Secretary-General to expedite the necessary arrangements so that the transfer could become effective by January 1993.

12. Ministers heard a report from the German Chief of Defence Staff on the meetings of Chiefs of Defence Staff. Ministers agreed that the Chiefs of Defence Staff should meet twice a year prior to the regular Ministerial Councils and on an ad hoc basis whenever necessary. Ministers also agreed that, following the transfer of the Council and Secretariat to Brussels, national delegations could be reinforced with military delegates to develop and provide advice for the Council, to introduce the views of the Chiefs of Defence Staff to the Planning Cell and to monitor the professional standards of the Planning Cell's work.

13. WEU Ministers welcomed the IEPG Defence Ministers' decision, at their Oslo meeting on 6 March 1992, to analyse the future role of the IEPG in the new European security architecture. This represents a positive development fully in line with the objective set by WEU member States in Maastricht further to examine enhanced cooperation in the field of armaments with the aim of creating a WEU European Armaments Agency. WEU Ministers propose that both WEU and IEPG experts analyse this issue in depth, carry out an initial examination of the role and functions of a possible European Armaments Agency and submit a report for consideration.

14. WEU Ministers welcomed the decision of Eurogroup Defence Ministers at their meeting in Brussels on 25 May 1992 to consider the possibility, among other options, and if the necessary preconditions are met, of transferring to WEU some or all of Eurogroup's present functions for which there is still a need.

15. Ministers noted with satisfaction the considerable progress which had been made in setting up the experimental WEU Satellite Centre in Torrejon (Spain), a concrete example of the strengthening of WEU's operational role, and they looked forward to the official inauguration which would take place later in the year. They also noted that the contract for the main system feasibility study had been awarded to a consortium of

firms from WEU member States led by a German firm.

Activities of Working Groups

16. In the field of verification, Ministers noted with satisfaction that a set of rules for the operation of multinational teams in CFE had been prepared in WEU and had subsequently been adopted in the Alliance. This represented the first example since the Maastricht Declaration of the introduction of joint positions agreed in WEU into the process of consultation in the Alliance.

17. Noting the progress which had been made so far, Ministers agreed in principle to a feasibility study to identify the most cost-effective means of implementing the Open Skies Treaty cooperatively among member States. They tasked the group of experts to agree assumptions for the study, to identify the options which merit further study and to consider the question of costs, with a view to taking a decision at their next ordinary meeting to proceed with the study. They stressed the readiness of WEU to cooperate with third parties at a later stage, and in this context welcomed the contacts which had taken place with other European allies, as well as with the Russian Federation. They agreed that experts should investigate the possibilities for intensified cooperation with the Russian Federation, which could include a joint feasibility study and/or a trial overflight.

18. Ministers reaffirmed the importance of the Mediterranean Sub-Group's work on security in the Mediterranean. They adopted terms of reference for the establishment by WEU of a gradual and phased dialogue with the Maghreb countries, taking into account the political developments both in these countries and in the region.

WEU Institute for Security Studies

19. Ministers noted with satisfaction the activities of the WEU Institute for Security Studies in Paris. Its publications, seminars and colloquia had

greatly contributed to deepening understanding for the ongoing development of a European security identity and to enhancing relations between WEU and other European countries.

II. ON STRENGTHENING WEU'S OPERATIONAL ROLE

1. In accordance with the decision contained in the Declaration of the member States of WEU at Maastricht on 10 December 1991 to develop WEU as the defence component of the European Union and as the means to strengthen the European pillar of the Atlantic Alliance, WEU member States have been examining and defining appropriate missions, structures and means covering, in particular, a WEU planning cell and military units answerable to WEU, in order to strengthen WEU's operational role.

2. WEU member States declare that they are prepared to make available military units from the whole spectrum of their conventional armed forces for military tasks conducted under the authority of WEU.

3. Decisions to use military units answerable to WEU will be taken by the WEU Council in accordance with the provisions of the UN Charter. Participation in specific operations will remain a sovereign decision of member States in accordance with national constitutions.

4. Apart from contributing to the common defence in accordance with Article 5 of the Washington Treaty and Article V of the modified Brussels Treaty respectively, military units of WEU member States, acting under the authority of WEU, could be employed for:

- humanitarian and rescue tasks;
- peacekeeping tasks;
- tasks of combat forces in crisis management, including peacemaking.

5. The planning and execution of these tasks will be fully compatible with the military dispositions necessary to ensure the collective defence of all Allies.

6. Military units will be drawn from the forces of WEU member States, including forces with NATO missions - in this case after consultation with NATO - and will be organized on a multinational and multi-service basis.

7. All WEU member States will soon designate which of their military units and headquarters they would be willing to make available to WEU for its various possible tasks. Where multinational formations drawn from the forces of WEU nations already exist or are planned, these units could be made available for use under the authority of WEU, with agreement of all participating nations.

8. WEU member States intend to develop and exercise the appropriate capabilities to enable the deployment of WEU military units by land, sea or air to accomplish these tasks.

9. A Planning Cell will be established on 1 October 1992, subject to practical considerations, under the authority of the Council. It will be located with the Secretariat-General in a suitable building in Brussels. The Council has today appointed Maj. Gen. Caltabiano (Italian Air Force) as its first Director. The Planning Cell will be responsible for:

- preparing contingency plans for the employment for forces under WEU auspices;
  - preparing recommendations for the necessary command, control and communication arrangements, including standing operating procedures for headquarters which might be selected;
- - keeping an updated list of units and combinations of units which might be allocated to WEU for

specific operations.

10. The Council of Ministers approved the terms of reference for the Planning Cell.

III. ON RELATIONS BETWEEN WEU AND THE OTHER EUROPEAN MEMBER STATES OF THE EUROPEAN UNION OR THE ATLANTIC ALLIANCE

A. Following the Declaration released in Maastricht on 10 December 1991 in connection with the Treaty on European Union, WEU Ministers recalled the fundamental principles on which relations between member States and associate member States should be based:

- settlement of their mutual differences by peaceful means, in accordance with the obligations resulting from the modified Brussels Treaty, the North Atlantic Treaty and the United Nations Charter, the commitments entered into under the terms of the Helsinki Final Act and the Paris Charter, and the other generally recognized principles and rules of international law.
- in their mutual relations, refraining from resorting to the threat or use of force, in accordance with the United Nations Charter.

They also stressed that thc security guarantees and defence commitments in the Treaties which bind the member States within Western European Union and which bind them within the Atlantic Alliance are mutually reinforcing and will not be invoked by those subscribing to Part III of the Petersberg Declaration in disputes between member States of either of the two organizations.

B. In their Maastricht Declaration of 10 December 1991, the member States of WEU proposed that States which are members of the European

Union be invited to accede to WEU on conditions to be agreed in accordance with Article XI of the modified Brussels Treaty, or to become observers if they so wished. Simultaneously, other European member States of NATO were invited to become associate members of WEU in a way which would give them a possibility of participating fully in the activities of WEU.

In accordance with Part III of the Petersberg Declaration, Ministers agreed that the following points should be made in extending the invitation to the countries interested in becoming members, observers or associate members:

Members:

Member States of the European Union which have accepted the invitation to accede to WEU undertake:

- to respect, in accordance with the principles and values adhered to by all WEU member States, the Brussels Treaty of 1948, modified on 23 October 1954, its Protocols and associated texts, and the agreements concluded among the member States pursuant to the Treaty,
- to note with approval the agreements, decisions and rules adopted in conformity with the Treaty, and the Declarations starting with the Rome Declaration of 27 October 1984,
- to develop WEU as the defence component of the European Union and as the means to strengthen the European pillar of the Atlantic Alliance in keeping with the obligation entered into on 10 December 1991 in the Declaration on the role of WEU and its relations with the European Union and with the Atlantic Alliance attached to the Treaty on European Union, and
- to accept in full the substance of Part III of the Petersberg Declaration which will form part of the Protocol of Accession.

Observers:

Member States of the European Union, which have accepted the invitation to become observers,

- may, although not being a party to the MBT, attend the meetings of the WEU Council without prejudice to the provisions laid down in Article VIII of the modified Brussels Treaty; at the request of a majority of the member States, or of half of the member States including the Presidency, presence at Council meetings may be restricted to full members;
- may be invited to meetings of working groups;
- may be invited, on request, to speak;
- will have the same rights and responsibilities as the full members for functions transferred to WEU from other fora and institutions to which they already belong.

Associate Members:

Other European member States of the Atlantic Alliance which have accepted the invitation to become associate members of WEU, although not being parties to the modified Brussels Treaty, may participate fully in the meetings of the WEU Council - without prejudice to the provisions laid down in Article VIII of the modified Brussels Treaty - of its working groups and of the subsidiary bodies, subject to the following provisions:

- at the request of a majority of the member States, or of half of the member States including the Presidency, participation may be restricted to full members;
- they will be able to be associated to the Planning Cell through a permanent liaison arrangement;
- they will have the same rights and responsibilities as the full members for functions transferred to WEU from other fora and institutions to which they already belong;
- they will have the right to speak but may not block a decision that is the subject of consensus among the member States;

- they may associate themselves with the decisions taken by member States; they will be able to participate in their implementation unless a majority of the member States, or half of the member States including the Presidency, decide otherwise;
- they will take part on the same basis as full members in WEU military operations to which they commit forces;
- they will accept in full the substance of Section A of Part III of the Petersberg Declaration which will form part of the association document;
- they will be connected to the member States' telecommunications system (WEUCOM) for messages concerning meetings and activities in which they participate;
- they will be asked to make a financial contribution to the Organization's budgets.

Space Activities

For practical reasons, space activities will be restricted to the present members until the end of the experimental phase of the Satellite Centre in 1995. During this phase the new members and associate members will be kept informed of WEU's space activities. Appropriate arrangements will be made for associate members to participate in subsequent space activities at the same time as decisions are taken on the continuation of such activities.

Mandate

C. Ministers mandated the Permanent Council to arrange for discussions to start with the States concerned.

Ministers confirmed their wish to conclude the necessary agreements before 31 December 1992.

INDEX

F

G

H

O

P

R

S

T